Program Evaluation

A Field Guide
for Administrators

Program Evaluation
A Field Guide
for Administrators

Robert L. Schalock
Hastings College and
Mid-Nebraska Mental Retardation Services, Inc.
Hastings, Nebraska

with

Craig V. D. Thornton
Mathematica Policy Research, Inc.
Princeton, New Jersey

Plenum Press • New York and London

Library of Congress Cataloging in Publication Data

Schalock, Robert L.
  Program evaluation: a field guide for administrators / Robert L. Schalock with
Craig V. D. Thornton.
      p.   cm.
  Includes bibliographical references and index.
  ISBN 0-306-42840-7
  1. Human services — United States — Evaluation. I. Thornton, Craig V. D. II. Title.
HV91.S277   1988   88-12608
361'.0068 — dc19   CIP

© 1988 Plenum Press, New York
A Division of Plenum Publishing Corporation
233 Spring Street, New York, N.Y. 10013

Printed in the United States of America

This book is dedicated to our wives, Susan and Kate, to program administrators who provided much of the material and all of the purpose for the book, and to the airports and telephone booths that made our collaboration possible.

Preface

This book is written to help human service program administrators either interpret or conduct program evaluations. Our intended audience includes administrators and those students being trained for careers in human services administration. Our focus is on persons interested in assessing programs in which people work with people to improve their condition.

The book's title, *Program Evaluation: A Field Guide for Administrators,* describes how we hope you use this book—as a tool. In writing the book, we have attempted to meet the needs of persons who have to conduct program evaluations as well as those who must use those evaluations. Hence, we have attempted to make the book "user friendly." You will find, for example, numerous guidelines, cautions, and specific suggestions. Use the book actively.

Our primary motive is to help administrators make better decisions. In fact, the primary reason for program evaluation is to help program administrators make good decisions. These decisions often must balance the goals of equity (or fairness in the distribution of goods and services among people in the economy), efficiency (obtaining the most output for the least resources), and political feasibility. Take, for example, the administrator who must decide between a new program favored by some of the program's constituents, and maintaining the status quo, which is favored by other constituents. The administrator's decision, which is made within the framework of equity, efficiency, and political feasibility, must also be made in an uncertain world whose resources of money, manpower, time, and experience are frequently changing. Our proposed approach to program evaluation guides analysts to select what information to get, and in what ways to obtain and use the information so as to reduce uncertainty and thereby make better decisions.

Recent years have seen a huge growth in the number of human service program evaluations, with the vast majority of these studies reporting positive findings. An administrator trying to sort through these evaluations is often faced with the task of determining not only the relative effectiveness of the various programs, but also the relative accuracy of the evaluations. The book's format and discussions are designed to provide the quick access to information about analytic techniques and interpretations that an administrator needs.

In the end, administrators must make decisions that reflect their assessment of the political and social implications of programs as well as of the objective evidence about program impacts. Furthermore, administrators will never have complete information and certainty; thus, decisions must be based on experience and value judgments in addition to evaluation information. The real benefit of the analytic procedures we present is to help the administrator organize and interpret available data. Analytic techniques cannot make decisions; they can only help smooth and clarify the decision process.

Program analysis therefore provides rules of evidence for organizing information to make judgments. In this regard, we provide more of an approach to decision making than a precise prescription. We do not ask that all programs be evaluated using large-scale, comprehensive, state-of-the-art methods, or that all aspects of the program be quantified. In most cases, such an intensive effort would be unwarranted, and in any event, there will always be some residual uncertainty due to the impossibility of obtaining information about all aspects of a program or course of action. What we do ask is that analysts be systematic in their collection and analysis of program data, and that their conclusions be based on thoughtful, careful, and honest analysis.

Research into the effectiveness of human service programs has a mixed record, at best. A few studies have shown the potential for program analysis by illuminating policy and helping us organize the available resources to best serve the needs of society. Nevertheless, many studies fail to realize their potential by using flawed methods and inadequate observation, overstating their conclusions, or unintelligibly presenting results. We hope that this book will serve to improve this situation by increasing awareness and understanding of analytic methods as they can be used in the human services field.

Our secondary motive for writing the book is to help administrators match resources to evaluation questions asked. Both authors are called on to provide technical assistance to programs that are attempting to conduct some type of program evaluation. What we find frequently is a misallocation of evaluation resources, wherein the program does not have adequate resources to answer large-scale evaluation questions that are frequently asked of it. A common example is a funding agency that requests an impact evaluation study, but adds only a token amount of funding for that evaluation. This situation leads frequently to both a bad evaluation and a frustrated administrator. Hence, our motive is to help guide the administrator in matching resources with evaluation questions, and in knowing which questions to go after if evaluation resources are limited.

In this regard, it should be noted that the tools and tasks should be matched. You would not go deep-sea fishing with a bamboo pole. If you only have the equipment for catching bass, you are better off to go to the lake where there are bass and fish there successfully and enjoyably. Similarly, program evaluation can be done successfully with limited resources, if the goals and tools are appropriately matched.

Writing a book such as this is both challenging and rewarding. Our own evaluation efforts and this book have benefited from the advice and experiences of many colleagues. In particular, we want to thank Charles Mallar, Peter Kemper, David Long, and other researchers at Mathematica Policy Research; human service program personnel and participants from throughout the country who have shared their evaluation data with us; and Gloria Mills, who has provided deeply appreciated technical support.

Robert L. Schalock
Craig V. D. Thornton

Hastings and Princeton

Contents

II. Process Analysis

4. Process Analysis from the Participant's Perspective

5. Process Analysis from the Program's Perspective

6. Analysis of Program Costs

III. Impact Analysis

7. Measuring Program Outcomes

8. Estimating Program Impacts

IV. Benefit–Cost Analysis

9. Our Approach to Benefit–Cost Analysis

10. Back-of-the-Envelope Benefit–Cost Analysis

V. Analysis to Action

11. How to Communicate Your Findings

12. How Am I Doing?

Program Evaluation

Program Analysis

This book is written to help human service program administrators be better producers and consumers of program evaluations, and therefore be able to answer the frequently asked question, "How am I doing?" We anticipate that reading the book and using it actively as a field guide will enable administrators to match resources to evaluation questions and thereby use program evaluation information to make better decisions. The primary purpose of Section I is to build the foundation for your role as either a producer or consumer of program evaluation, so that when you read Sections II–IV you will be in a better position to follow the evaluation process and to draw on the various types of evaluation activities.

Chapter 1 discusses decision making and evaluation in program administration. It is organized around four areas that set the stage for the remainder of the book. We examine programs, perspectives, and the decision environment; some of the administrator's decision-making dilemmas regarding evaluation; programmatic decisions and evaluations; and the three types of program analysis—process, impact, and benefit–cost—that we will cover in the book. Throughout the chapter, we present guidelines and cautions that should be helpful in either your producer or consumer role. We also introduce you to evaluation studies that will be used throughout the book to provide examples of significant program evaluation issues and techniques.

Chapter 2 permits you to play the producer's role by having you describe your program's purpose, population, and process. As a producer, you are asked to answer five questions frequently asked in program evaluation, including the problems addressed, the persons served, what services or interventions are provided, the expected outcomes, and the justification for why you think your program is going to work. The chapter also proposes a test that any evaluation report produced should be able to pass. It's called the "mother-in-law test," and it asks, can a reader who is willing and careful, but also nontechnical and somewhat skeptical, understand what you are saying? This is not only an issue of presentation but also one of logic and common sense. We recognize that in some cases complex procedures will be needed, but a good report will present the basic underlying logic in a way that makes sense to an interested, yet non-technical audience. Just as in the case of a stereotypical mother-in-law, such an audience is more likely to be convinced by a straightforward logical explanation than by appeals to esoteric theories and techniques. Also, by stepping back from your program and its evaluation and trying to sort out what you are trying to accomplish, you can avoid many of the evaluation pitfalls produced by over-familiarity with the program and its basic assumptions.

Chapter 3 is a critical chapter and in a sense represents the book's fulcrum. The chapter includes a discussion of a number of guidelines governing program evaluation that we feel apply to any type or intensity of program analysis. The rules are presented within an analytical framework that stresses three evaluation phases: the setup, marshalling the evidence, and interpreting the findings. The setup defines the evaluation goal, context, and rationale that links the services provided to the expected outcomes. Marshalling the evidence requires collecting evidence to support the setup and to justify the contention that the intervention produced the desired effects. Interpreting the findings involves addressing the correspondence between the setup and the evidence, determining whether the effects can be attributed to the intervention, and evaluating the results in light of the program's context and limitations. Chapter 3 should provide you with the conceptual framework to begin your active role as a producer or consumer of process, impact, or benefit–cost analyses, which are presented in Sections II–IV.

Throughout Section I, we stress that program evaluation is undertaken to facilitate decision making. Decisions are choices such as between two programs, between a new program and the status quo, or between a new and an old policy. In each of these cases, a comparison is being made. Structured comparisons are at the heart of evaluation. If there was no comparison, there would be no choice or decision and therefore no need for evaluation. Thus, as the first step, the evaluation must specify the program or policy being evaluated and the program or option with which it will be compared. This specification should include information on such factors as the persons being served, the services being offered, the environment in which the program or policy will operate, and the environment in which it will be compared. These two alternatives—the program and the comparison situation—define the scope and ultimately the results of the evaluation. All components of the evaluation, including interpretation of the findings, must be undertaken in relation to these two alternatives. In reading these three chapters, we trust that you will become more familiar with program evaluation and more sensitive to your role as a producer or consumer of it.

1

Decision Making and Evaluation in Program Administration

I. Overview

Program administrators are decision makers. Others may set program goals, gather information, carry out policy, or deliver services, but it is the administrator who must make the decisions as to who will be served and how they will be recruited, screened, and enrolled; what services will be provided over what time period and through what means; and who will staff the program and provide specific types of services. These are the types of decisions administrators make and remake as programs are initiated and then operated over time.

These decisions require evaluation, which is the systematic collection and analysis of information about alternatives. The evaluations may be informal and quick, or they may be complex, highly structured efforts. When a decision involves few resources or can easily be researched, little evaluation is required. Decisions about where to lunch, where to purchase office supplies, or whether to include someone on a mailing list are often made with little systematic evaluation of the options. In contrast, office location, staff hiring, program development, and program funding involve decisions that have far-reaching implications and so are generally made on the basis of more careful evaluations.

These decisions and evaluations are inescapable parts of all programs and, in fact, all life. Thus, it is worthwhile to consider the efficiency of decision making and evaluation. This involves assessing the appropriate types and intensities of evaluation efforts, their accuracy, and their limitations. It involves developing guidelines and rules of thumb to assist in evaluation. It also involves examining specific evaluation applications, since the trade-offs inherent in the day-to-day practice of evaluation are not easily seen in the context of actual evaluations. Throughout this and subsequent chapters, we review the general guidelines and rules of evidence that underlie all evaluations and examine the methods for using these to assess the types of decisions that arise in human service programs.

At this point, you might be wondering, why do I need another book on program evaluation? Aren't there enough? We don't think so, for we have observed an unmet need which this book attempts to fill. The need is that program administrators are seldom trained in program evaluation, and yet they are frequently expected to either conduct evaluation studies or interpret others' evaluations. Thus, at the outset we make a distinction between administrators as producers and consumers of evaluation. As producers, administrators need to know the analytic techniques to use, given the questions asked and the available resources. As consumers, administrators need to know if they can act on the reported findings of other studies. In reference to both roles, we review the range of evaluation activities, introduce you to certain rules of evidence, and suggest approaches for getting the most out of your own evaluations as well as those done by others.

There is considerable program evaluation literature on how to make big decisions, and little literature is needed with regard to small decisions such as buying a new typewriter or this book. In this book we focus on the middle, wherein the administrator doesn't have the resources to do large program evaluations, but the decision is important enough to have systematic review and thought. Examples of decisions an administrator might face include:

- Whether to fund a sheltered workshop, or undertake a supported employment initiative that would switch state resources from one program model to another.
- Whether to reduce the dosage of medication in patients within a mental hospital as suggested by advocates and accreditation groups, and if this is done, what will be the effects.
- What evaluation activities are appropriate to include in a small demonstration program with a limited budget.
- Whether to attempt to convince others that a program looks promising and may be feasible for wider implementation.
- Whether to staff a program with bachelors-level rather than masters-level personnel.

Administrators are faced constantly with these types of decisions, which they often must make under substantial time pressures. As an administrator makes the day-to-day decisions relating to program operations, reports to funders, coordinates staffing and supervision and reacts to the other exigencies of human service programs, the thought of undertaking a program evaluation must appear particularly daunting. This is true even if the administrator can hire an evaluation specialist to conduct the work, since the administrator must still pose the appropriate questions to the evaluator and assess the resulting evaluation to determine its accuracy and relevance.

In this book, we try to provide guidance for administrators to help them utilize the power of evaluation methods without the necessity of becoming experts themselves. Our approach is to provide a general framework for evaluation, along with some rules of thumb and guidelines. Our intent is to help administrators to focus their evaluation efforts, to match their resources with their evaluation needs, and to provide a basis for them to judge more efficiently the merits of evaluations conducted by others.

The general framework provided in this book underlies all good evaluation efforts, regardless of their complexity or scope. The basic approach we present is equally valid for a large-scale rigorous evaluation and for a quick informal weighing of available evidence. The difference in these evaluations is not in their basic structure, but in the degree of certainty produced and the complexity of the questions addressed. This book will prepare readers to interpret all types of evaluations and to conduct a wide range of modest evaluations. We have not attempted to summarize all the sampling, statistical, data collection, and interpretation techniques that one uses in evaluations today. Rather we guide readers as to what to look for in an evaluation; what questions to ask of evaluators before, during, and after the evaluation; and how to focus your own program's resources and information to illuminate program decision making. Rules of thumb presented herein are just that, general approximations that are useful for designing and interpreting evaluations. These rules serve as useful guides, but are not inviolate. However, when these rules are not followed, it is reasonable for an administrator or other consumer of evaluations to ask for a clear reason why. In these cases, the references provided in each chapter will be useful in providing additional specifics about the general guidelines and procedures.

The purpose of Chapter 1 is to outline our approach to program evaluation and the structure of the book. Our intent throughout the book is to present guidelines that should be helpful in either your producer or consumer role. Analogously, wherever appropriate we present cautions that we hope will prevent unwarranted, unnecessary, or involved procedures or interpretations. Our cautions are very evident, for they are highlighted within the text. These cautions are necessary because program analysis involves technical skills and strategies that one cannot overlook. We anticipate that our guidelines and suggestions will be helpful and will result in better use of evaluation data and activities. But we also feel strongly about cautioning the reader at critical points about highly technical or debatable issues.

The chapter is organized around four areas that set the stage for the remainder of the book. These include (1) programs, perspectives, and the decision environments; (2) program administrator's decision-making dilemmas; (3) programmatic decisions and evaluations; and (4) types of program analysis. Throughout this and subsequent chapters, we emphasize that:

• There is uncertainty in all we do, including program evaluation.

- Evaluation is a continuous, dynamic process, and the issues are always changing.
- Evaluation involves clear, logical thinking.
- Evaluation techniques cannot make decisions; they can only inform, smooth, and clarify the decision process.
- Regardless of the scale of evaluation, one must be logical and follow general scientific methods.

II. Programs, Perspectives, and the Decision Environments

Decisions are not made in a vacuum. They are made in society, where administrators face many pressures. The first step in evaluation is to understand the decisions being made and the context in which they are made and will be judged. In society, there are necessarily many competing interests that result from diversity within the population. Figure 1.1 shows some key aspects of the system and some conflicting incentives that logically arise in society. These groupings are denoted in Figure 1.1 by boxes around different groups of individuals. Individuals form into a vast number of interest groups, with any particu-

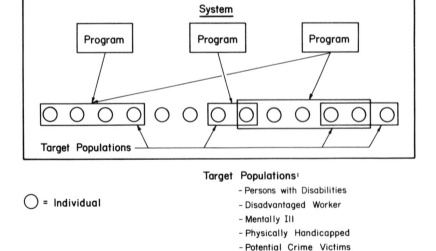

Target Populations:
- Persons with Disabilities
- Disadvantaged Worker
- Mentally Ill
- Physically Handicapped
- Potential Crime Victims
- Elderly

◯ = Individual

Figure 1.1. A program and its constituency.

lar person falling into many interest groups. These groups include organizations such as labor unions, employer groups, persons with disabilities, older persons, unemployed persons, individuals in industries threatened by foreign competition, potential victims of crime, persons living in specific regions, and individuals with similar religious or philosophical orientations. Social programs are created to serve these interest groups, and a specific program may serve several interest groups. The "system" encompasses all these individuals and programs.

Evaluation tends to use the *efficiency criterion* of society: does a program or policy contribute to the overall value of social goods and services? But there are other legitimate criteria that should be recognized, such as self-interest of program staff, interest groups, and participants. Therefore, when conducting an evaluation, it is useful to consider this "system" to assess the overall issues you want to address, the interests of the groups involved, and the other criteria that might be used.

One such criterion is *equity,* which refers to the fairness in the distribution of goods and services among people in the economy. Each group shown in Figure 1.1 may differ in its definition of equity. In program evaluation, one can often say something about efficiency, since more goods and services are generally considered desirable. Furthermore, one can often describe effects on equity, but one will have limited (at best) ability to judge or value changes in equity.

Thus, we suggest that you do not separate evaluation activities from the system within which they occur. That system includes individuals; the program they demand and create; and the government system that organizes, operates, and monitors the program(s) to balance the interests of various groups. One should always remember that individuals wear two hats: they pay for government services and they consume those services. Hence, interests vary depending on which hat is being worn with respect to a particular issue.

For example, program evaluation done from the perspective of individual service consumers asks whether the interest group served by a program is being helped by the program. Can they be expected to enroll or participate in the program voluntarily? Will they benefit from their participation? Individuals can generally be assumed to be motivated by their own self-interest and will attempt to receive services accordingly. They will be interested in programs that can best help them achieve their goals. Of course, individuals are also likely to include the general welfare of other individuals in their set of goals, and thus they can be expected to support some programs aimed at improving the situation of others.

Individuals as funders of the system will be interested in an evaluation of how well the system as a whole satisfies their aggregate needs. In particular, they are interested in whether the set of existing programs is the best one. They will ask whether there is another set of programs that can achieve better results or the same results at a lower net cost. These are the types of questions program evaluation asks when examining the system level.

The mechanism for satisfying individual needs and for translating social demands into actions is the program. Programs face an intriguing set of incentives, since a specific program need not be interested in the overall performance of all programs. Rather, a program (or more precisely its staff and supporters) will be interested in its own survival and the continued satisfaction of the demands that led to its creation and the continued service to the interest groups that support it. This interest can lead the program to seek to improve its operations, thereby better satisfying social needs and increasing demand and support for the program. This interest can also lead to attempts to disseminate only positive information about the program and to mask any problems. The program analyst must recognize these sometimes conflicting implications.

Thus, many different persons or constituents are interested in program evaluation: the participants (or more broadly the target population), program staff, funders (foundation, legislature, department, or agency), and society in general. The types of questions they ask, and the decisions required of a program administrator, may be very different.

III. Program Administrator's Decision-Making Dilemmas

Program analysis involves being sensitive to the program administrator's perspective and potential dilemmas. No program administrator wants to have a bad program, to be considered a loser, or to have the program terminated. Our experiences have been that most program administrators face a number of potential dilemmas.

An important dilemma is caused by trying to keep human service programs in step with changes in technology, social needs, and prevailing attitudes. Programs may be established for one purpose and then have other goals added over time as society looks for new or better ways to meet the needs and demands of its members. Such change can bring an administrator to a situation in which his or her program is being evaluated with criteria that were not considered when the program was established. This updating is not necessarily incorrect, since programs should be responsive to changing social situations. It does, nevertheless, create pressures for the administrator, and these pressures should be considered in the evaluation.

For example, this situation has arisen for operators of sheltered workshops for persons with mental retardation. In the 1950s and 1960s, these programs were considered at the leading edge of service for such persons. The programs provided a stable environment where persons with mental retardation could work, meet friends, and receive, to some extent, training that might enable them to move on to a job in the regular labor market. In the late 1970s and the 1980s, training methods and social attitudes and expectations had changed to the extent that these services were no longer thought to be appropriate for many persons with moderate or mild retardation. Instead, the new methods stressed training these persons directly on regular jobs in the economy rather than in segregated workshops. This change left some workshop operators with a feeling of betrayal.

Similar scenarios are found in the human service areas of mental health (community mental health centers versus mental hospitals), education (classroom versus community-referenced instruction), and aging (nursing homes versus community-based care). Evaluators must recognize these types of issues in the historical development of the programs being studied. The evaluation may still use criteria that reflect the current social expectations rather than those that gave rise to the program, but sensitivity to changing criteria can help the evaluator to understand the perspectives of program operators and advocates for change. Sensitivity to these issues can also improve the presentation and ultimate acceptance of evaluation results and increase the chances of developing effective improvements in program operation.

A second potential dilemma arises when a program's goals conflict with, or create conflicts when combined with, another program. Programmatic goals are not alwasys consistent with one another. Examples include tobacco subsidies and health warnings on tobacco products, and sheltered workshops with their emphasis on stability and retaining high producers versus a transitional program the goal of which is to place participants into competitive employment. Other examples relate to a number of public programs that have been implemented to insure income support and medical care for persons who are handicapped physically, mentally, or economically. These programs often create disincentives for program participants to work by offering potentially greater income security than could be obtained in regular employment to persons who continue as beneficiaries of these programs (Conley, Noble, & Elder, 1986). Such disincentives inhibit the success of training programs that seek to place these persons into jobs.

The issue for administrators and evaluators is the impact other programs may have on the performance of the program under study. At the least, there must be a description of these other programs and an assessment of the potential for conflicts. Generally, a program will be evaluated in its current environment with all the resulting interactions. This reflects the fact that a single program administrator cannot resolve all the potential conflicts between programs and so must assess his or her program as it operates in the real world. However, more system-wide evaluations may seek to assess how a program's performance would be improved by closer coordination of program goals. Such an evaluation can be instrumental in bringing about structural change in the human-service system. The evaluator must consider which perspective is appropriate for the decisions at hand and then must pursue the evaluation from that perspective.

A third potential dilemma results from conflicts between client level goals and program goals. For example, consider the financial risks to a sheltered workshop of moving a productive worker into a regular job in the community. The workshop loses a valuable employee and generally must fill the vacated position with someone who will be less productive, at least initially. This forces the workshop to choose between its need for good workers who can ensure that the workshop will meet its production needs and the goal to move the worker into a more independent and more integrated job. This type of conflict puts many programs in a catch-22 situation during times of increasing demands for program accountability but of shrinking dollars and limited resources. Socially, there are also risks to the agency. What if a participant within a mental health, criminal justice, mental retardation, or halfway house for chemical dependency "goes astray" or breaks the law? That one client can put the entire agency's or program's continued operation in serious jeopardy. The net result is that program administrators (and indeed programs) become conservative, which frequently leads to a bias toward the status quo.

The challenge posed by all these dilemmas is to define exactly what is being evaluated, the context within which the program operates, and the appropriate criteria and perspectives for assessing performance. Many administrators who face daily these dilemmas fear that evaluators will miss the inherent conflicts and will develop inaccurate or inappropriate criteria. This fear is frequently exacerbated by the demands of funders who often base their judgments on narrowly defined performance measures (such as cost per person placed on a job). Evaluators must be sensitive to these concerns and work to incorporate them as appropriate into their work.

IV. Programmatic Decisions and Evaluations

We expect that after you have read this book you will be a better consumer and producer of program evaluations. As a consumer of evaluation, you will know if you can act on the evaluation study you have just read. We also expect that you will gain the perspective of an evaluation insider in the sense that you will be able to distinguish those evaluations that are well done. A problem faced by all persons who are new to the field is sorting through all the studies that report evaluation findings and determining whom to believe. An administrator or staff person who is asked to search for evidence on a particular program or policy may find dozens of articles, books, reports, and opinions about the topic, particularly if the issue is controversial and of interest to several advocate groups. The insider ability we want to instill is how to sort through those various items to find the best evidence, to eliminate the flawed studies, and to identify those that provide accurate and useful information about the topic in which you are interested. Because of the uncertainty inherent in social policy and program evaluation, there may often be conflicting evidence and uncertainty even in the best studies. The insider will recognize this and use the available information to identify the key areas of uncertainty and the areas where more information (or assumptions) will be needed to make the required decisions.

As a consumer of evaluations, it is necessary to ask of each evaluation: What is the level of certainty in the findings? How likely is it that if the study was replicated, the same results would be obtained? Is the set of experiences on which the study is based sufficiently well documented and broad enough to support the conclusions? Finally, are the comparisons made in the evaluation relevant to the decisions that must be made, and is the level of certainty sufficient for action to be taken on the evidence reported in the evaluation? In that regard, let's look at the issues of uncertainty, applicability, and reasonableness.

1. *Uncertainty*. The first thing that both administrators and evaluators face is the considerable uncertainty involved in any decision. Causes of uncertainty relate to sampling problems, measurement error, and multiple outcomes. In reference to sampling problems, there is considerable variation in human activities, interests, attitudes, and abilities. In program evaluation, we observe the activities of only a *sample* of all the persons who might use a program or be influenced by a policy. As a result, there is uncertainty because we do not know if the next sample of persons will act as our original sample did. In addition, sample uncertainty is due to the fact that the intervention may not be the same over time. The environment may change, or the staff may modify what they do. In either case. the results may change, and thereby we have uncertainty.

Information about program effects requires that we measure the activities of persons affected by the program. There is error inherent in all measurement. Age can be measured accurately (subject to reporting error), but intellectual ability, as an example, is very hard to measure accurately. Thus, measurement error also creates uncertainty.

And third, uncertainty is related to the fact that human service programs have multiple outcomes, and there is uncertainty about how important different program effects are. It's similar to the acid rain debate: acid rain policy requires that job effects be weighed against health and environmental effects. Frequently, we do not know how to weight these multiple outcomes, but since we still have to make decisions, some process must be adopted to resolve the uncertainty.

In reference to these three sources of uncertainty, we can deal with sampling problems and measurement error better than with multiple outcomes. With sampling problems, uncertainty is generally diminished as one uses larger and better designed samples. Measurement error is reduced through controlling extraneous factors and using measures that are reliable and valid. It's more difficult to get a handle on the issue of multiple outcomes, but later we suggest the use of sensitivity tests and other procedures that reduce the amount of uncertainty in multiple outcomes.

2. *Applicability*. The question you need to ask concerning applicability is

what is the comparison made in the evaluation and what is the comparison you are interested in? (That is, what comparison forces the decision maker?) Do they differ with respect to the population being studied, the time period, the environment, the specific intervention or policy, or other relevant factors? Evaluation results are applicable primarily to those programs or situations that are similar to the author's program; therefore, as a consumer, you will need to determine how similar the reported participants, services provided, and environment within which the reported program operates are to those for the program about which you must decide.

3. *Reasonableness.* What is reasonable? You will never have complete certainty or complete applicability. What is a reasonable match will depend on the level of certainty you feel you need for your decision. Informal or easily changed decisions can be based on casual observation, but more important decisions may require a closer match and more certainty. At the extreme, you may need to do a formal demonstration in which the specific policy is tested in a natural environment and subject to a rigorous evaluation. But this will be costly and time consuming.

Our goal throughout the book is to help you understand the sources of uncertainty and the trade-offs made in using empirical evidence to make policy decisions. The importance of this goal is reflected in a recently published article entitled, "A Critical Analysis of the Diets of Chronic Juvenile Offenders" (Schauss & Simonsen, 1979). These authors report a statistically significant difference in the diets (particularly milk consumption) between delinquents and nondelinquents. Such a finding could be interpreted cautiously (as the authors do), or it could be used to support the hypothesis that milk causes delinquency. A reader of this article should immediately ask a number of questions regarding uncertainty, applicability, and reasonableness, including: How many subjects were employed, and how were they selected?; How reliable and valid were the measures used?; Were proper comparisons made, and were the comparisons significant?; Were the participants of the study similar to those in my program so that I might apply the results? For example, if you were an administrator of a mental health facility, would you stop serving milk to your clientele? Probably not, because of the questions you have about the study's uncertainty, applicability, and reasonableness. These concerns and other questions are important to raise as a consumer of evaluation, and throughout the book we provide rules of evidence and other guidelines to help you decide whether you can act on the results presented.

Often the easiest and most accessible means of making a decision is to see what other persons in similar circumstances have decided. Information about these decisions is available in the published reports and evaluations of other programs. As a result, the administrator is quickly faced with a decision about whether the other studies are reasonable bases for making decisions about his or her program. The administrator will assess the reasonableness by determining the quality of the reported analysis and the extent to which the reported situation resembles his or her own. Thus, our first guideline relates to the reader as an evaluation consumer. It reflects the uncertainty, applicability, and reasonableness issues that you must consider in making programmatic decisions based on evaluations.

> *Guideline 1.* As a consumer, the major question you want to pose in reading evaluation is the following: Is there enough uncertainty and applicability to my situation to act on the evaluator's results?

We anticipate that many of our readers will also be evaluation producers. Our expectation here is that you will want to conduct reasonable program evaluation activities in which the questions you ask are well matched with your available resources, and that your conclusions follow logically from what you observe. This leads to our second guideline, which pertains to the producer role and the importance of logical thought; that is, regardless of the size of your evaluation, you must follow the scientific method and use logical thought as summarized in the following guideline.

> *Guideline 2.* As a producer, you want to (1) define carefully the question you want to answer, (2) collect information about that question, and (3) interpret that information, making sure your conclusions follow logically from what you observe.

An example of our second guideline is an unpublished study in which the administrator of a mental hospital needed to know the effects of psychotropic medication reduction in a group of schizophrenic or organic brain syndrome patients. The question was asked because the psychiatric staff had begun reducing medication levels following an accreditation survey that suggested many patients seemed ''overly medicated.'' Ward staff, on the other hand, were voicing complaints of increased staff injuries and restraint usage in those patients whose medication had been reduced. Thus, the administrator set up a small evaluation study to answer the question, What are the effects of reduced psychotropic medication?

A reasonable amount of information had to be collected to answer the administrator's question. The more important data elements included drug levels, before the study and at various points thereafter, of patients for whom medication was increased, decreased, or kept the same; a standardized measure of behavioral change that would indicate whether the patient got better, worse, or stayed the same following drug dosage change; baseline and subsequent rates of staff injuries and restraint usage associated with patients whose medication was changed; and consistent time periods at which medication level, behavioral change indicators, restraint usage, and staff injury data could be compared.

The data, which were collected for a 1-year period, showed some interesting results, including the following: (1) behavior stayed the same in 73% of the patients regardless of whether medication levels increased, decreased, or stayed the same; (2) restraint usage did not increase in those patients whose medication was reduced; and (3) there was no increase over baseline in the number of staff injuries attributed to the patients whose medication was reduced 25% or more.

This was not a big study, but it provided the administrator with helpful information and allowed him to make a decision about the effects of reduced psychotropic medication. It also represented a reasonable trade-off between the information he needed and the costs involved in obtaining it. But it is also important to point out that even though the administrator followed Guideline 2, there was still some uncertainty in the evaluation's results.

Whether you are an evaluation producer or consumer, you will need information to make decisions. This information will have a degree of uncertainty, applicability, reasonableness, timeliness, and cost. As a decision maker, you must decide about the trade-offs between the information you need and the resources involved in obtaining it. You have a number of options as to where to get the information, including published program evaluation studies, or reports from similar programs or from your own evaluation. And furthermore, as the nature of your decisions change, you may need to change the type of evaluation you conduct. To illustrate this point, we would like to conclude this section by discussing five evaluation studies that involved three different types of decisions. For expository purposes, we suggest that decisions frequently need to be made regarding (1) a program's feasibility, (2) the demonstration of a program's generalizability to a larger population or environment, or (3) how well large, ongoing programs are doing. These different types of decisions will be related to three levels of program development.

A. Feasibility Stage

The feasibility stage of program development and evaluation focuses on demonstrating that the program is promising and/or feasible. The task is to convince others of the merits and potential of the new approach. Therefore, programs tend to be small and program evaluation activities center on documenting the program and its results.

A good example of this type of evaluation was recently reported by Mank, Rhodes, and Bellamy (1986), who evaluated an enclave model as one option for providing supported employment. As used by these investigators, an *enclave* is a group of persons with disabilities who are trained and supervised among nonhandicapped workers in an industry or business. The enclave whose feasibility was evaluated was housed within a host electronics company. A nonprofit organization funded by state service agencies provided supervision and support to the six individuals, labeled as severely handicapped, who were employed in a production line within the final assembly area of the plant. Ongoing supervision and training were provided by the lead supervisor and a model worker provided by the company.

After describing the program, the authors present preliminary data summarizing the first 2 years of the model's operation. Generally speaking, participants experienced increased wages compared with their preenrollment status, integration with coworkers, and chances of being hired by the company as a regular employee as they reached 65% productivity. Additionally, the cost data suggested a dramatic decrease in costs to the public sector over time.

The small number of persons observed and the program's use of only one work site limit the extent to which these results can be generalized. Nevertheless, by demonstrating that this enclave approach was feasible for persons with severe handicaps, this study provided a basis for further development of employment programs for these individuals. Furthermore, the careful documentation of procedures provided a basis for others to replicate the program and thereby build the larger set of experiences needed to assess the replicability and generalizability of this enclave program.

Feasibility evaluation data such as the above are very important to administrators and policymakers alike. Many administrators have good ideas and have implemented programs based on those ideas. In the 1960s, there was a huge growth of human service programs and expenditures which created the need for decisions and good data. As expenditures have tightened up, this type of evaluation is even more valuable. Thus, we devote Section II of the book to process analysis, which focuses on describing and collecting information about your program that will help you justify its feasibility and potential.

Once the feasibility of a program model has been shown, the focus of evaluation shifts to a larger scale in which significant questions about resource allocation and public policy arise. Thus, evaluation activities and appropriate techniques also shift because of the need for more certainty. Making these decisions will require larger studies involving more sophisticated techniques and resources. We realize that most readers will not be producers of demonstration-stage evaluation studies; but as consumers of these studies, it is important to be aware of their certainty and applicability (Guideline 1).

B. Demonstration Stage

At the demonstration stage, one needs to prove the program model on a larger scale. Thus, one is involved in evaluating data sets from multiple sites and using rigorous methodological and statistical techniques. At this point, we would like to introduce two large evaluation studies that are used throughout the book to provide examples, continuity, guidelines, and cautions to the reader. Both involved demonstration-stage evaluation, and both represent very large evaluation studies.

The first study is the National Long Term Care (LTC) demonstration that was funded by the U.S. Department of Health and Human Services to test two channeling models for organizing community care for the elderly (Kemper, 1986). The decision faced in the LTC demonstration was whether money could be saved by expanding the use of community-based long-term care services as an alternative to nursing-home care.

Two channeling models were tested. Both offered individuals who were at risk of institutionalization a systematic assessment of their needs and ongoing case management to arrange and monitor the provision of services. The models differed with respect to how community services were provided to clients. One model, the basic case management model, managed services that were available to clients in the community and added a modest amount of funding for purchasing services that were unavailable through other sources. The second model, the financial control model, expanded the range and availability of publicly financed services but, at the same time, instituted cost control features that placed a cap on average and per-client expenditures. The overall evaluation, which surveyed persons who participated in one of the two demonstration models or were in a control group, was designed to determine (1) the impact of the two models on the utilization of services and informal care givers, (2) the impact on client well-being (which is defined in a subsequent chapter), (3) the feasibility of implementing future channeling-type programs, and (4) the cost-effectiveness of the channeling concept.

The second study is the Structured Training and Employment Transitional Services (STETS) demonstration that was funded by the U.S. Department of Labor and administered by Manpower Demonstration Resource Corporation. This demonstration was undertaken to determine the effectiveness of transitional employment programs in integrating mentally retarded young adults into the economic and social mainstream of society (Kerachsky, Thornton, Bloomenthal, Maynard, & Stephens, 1985). The STETS program model was designed specifically to serve the needs of 18- to 24-year-olds whose IQ scores ranged between 40 and 80, who had no work-disabling secondary handicaps that precluded work, and who had limited prior work experience. It encompassed three phases of activity, including (1) initial training and support services that were generally provided in a low-stress work environment and which could include up to 500 hours of paid employment; (2) a period of training on potentially permanent jobs in local firms and agencies; and (3) follow-up services to help workers retain their unsubsidized, competitive employment. The research sample consisted of 437 individuals (226 experimentals and 211 control group members) residing in five U.S. cities. The STETS research plan was designed to address five basic questions: (1) does STETS improve the labor-market performance of participants; (2) does STETS participation help individuals lead more normal lifestyles; (3) in what ways do the characteristics and experiences of participants or of the program influence the effectiveness of STETS; (4) does STETS affect the use of alternative programs by participants; and (5) do the benefits of STETS exceed the costs?

These two studies, which are described in more complete detail in subsequent chapters, raise different evaluation questions than we saw with feasibility-stage evaluations. Note terms such as *comparison groups, costs, benefits,* and *impacts.* Section II through IV discuss these concepts in detail and present both consumer and producer guidelines.

One may occasionally be in a position to make decisions about demonstration-stage programs and yet not have the resources to engage in a large, multi-million dollar demonstration project. This was the case with Wehman and his associates (Hill, Hill, Wehman, & Banks, 1985) at The Research and Training Center at Virginia Commonwealth University. Since the fall of 1978, 155 persons with mental retardation have been placed into part- and full-time competitive jobs through the efforts of university-based federal grant support staff and local agency staff. Despite the lack of a comparison group, the investigators were interested in estimating the impacts of their program, under the assumption that had participants not enrolled in the program they would have continued in the activities they had prior to enrollment.

The analysis addressed questions about the potential of supported employment and focused on group benefits versus cost data from three perspectives: the society, the taxpayer, and the individual participant (Thornton, 1984). Measured benefit categories included monetary outcomes (earnings, fringe benefits, taxes paid), use of alternative programs (activity centers, workshops), and government subsidy costs (SSI, health care). Cost categories included operational costs, decreased workshop earnings, decreased subsidy (SSI and increased health costs), and taxes paid. In the analysis the actual or estimated monetary benefits and costs incurred due to intervention from the view of society, the taxpayer, and the participant were obtained from either participant records or published data from the U.S. Department of Labor and other sources.

Although the specific details of such a benefit–cost analysis are presented in Section IV, the point we wish to make here is that it is okay to attempt *limited* post hoc demonstration-stage evaluations; but if you do, realize that these evaluations represent very crude estimation procedures, and ultimately are not likely to withstand rigorous analytical scrutiny. Additionally, you need to test the sensitivity of the findings of the assumptions by recomputing the values using different cost estimates. Hence, a caution.

Caution. If you estimate what would have happened to participants had they not enrolled in the program, relying on a general knowledge about the average outcomes of nonparticipants or on the knowledge of preenrollment status, you need to realize that these methods represent very crude estimation procedures and ultimately are likely to produce uncertain estimates that often will not withstand rigorous analytical scrutiny.

In summary, the demonstration stage of development involves more sophisticated evaluation techniques that focus on demonstrating the program's costs, benefits, and impacts on a large scale. The third stage of program development involves an ongoing program with evaluation that focuses on determining how well the program is doing. It's to the third stage that we now turn.

C. Ongoing Stage

Many administrators are associated with large, ongoing programs within the broad service areas of mental health, education, labor, special education, corrections, developmental disabilities, or senior services. Policymakers frequently want to know how well these programs are doing. Their questions and related decisions generally center around the issue of resource allocation, program changes, effectiveness, and efficiency. Thus one sees numerous large, ongoing-stage program evaluations in the literature. We would like to use one such evaluation, that of evaluating the benefits and costs of the Job Corps (Long, Mallar, & Thornton, 1981), to demonstrate program evaluation activities associated with this stage of program development.

Job Corps provides a comprehensive set of services to over 38,000 disadvantaged youths annually. These services, primarily vocational skills training, basic education, and health care, are typically provided in residential centers and are intended to enhance the employability of participants. The program is expected to lead to increased earnings, reductions in corps members' dependence on public assistance, mitigation of their antisocial behavior, and corps members' improved health and quality of life.

In the evaluation study, the various effects of Job Corps were estimated using data collected in periodic interviews with corps members and a comparison group of similar youths who were never enrolled in Job Corps. A baseline survey of the two groups was administered in May 1977, and follow-up surveys were conducted over the next four years. Altogether, baseline and some follow-up data were collected for approximately 5200 youths. Statistical techniques, controlling for both observed and unobserved differences between corps members and youths in the comparison sample, were used with these interview data to estimate program effects during the period covered by the interview.

For an ongoing program like the Job Corps, the evaluation issue was whether the program was producing its desired effects. Of course, such a program might also be interested in ways to improve service delivery or in whether the program should be expanded or contracted. These latter issues might entail feasibility studies or demonstration efforts. Thus, administrators of ongoing programs may use several modes of evaluation, but the need to keep the program operating smoothly may constrain the options available to the evaluator.

D. Conclusion

Now, we don't expect you to conduct a 5-year survey of thousands of persons as was done for the Job Corps; but we do hope that your reading this book will help you know what to look for in the evaluation and to feel uncomfortable if key features are not there. Also, as you attempt to improve your program, you will face decisions and may be presented with evaluations proving the effectiveness of various alternatives. We anticipate that having read the book will allow you to interpret these evaluations and to be a good enough producer to set up a small feasibility-stage evaluation if needed.

And that brings us to Guideline 3, which is based on our discussion of the decisions that need to be made, the stages of program development, and the appropriate evaluation focus.

Guideline 3.   The evaluation focus must match the decisions that need to be made.

Given the previous discussion, the guideline should be self-explanatory. There are, however, two corollaries that we would like to add. First, a few sound answers are better than a lot of conjecture; and second, to be an informed consumer, you need to be a skillful producer. Thus, regardless of the complexity of the evaluations described throughout the book, we will tell you a little about producing effective evaluations so that you will be a better consumer. Three program analysis techniques related to the guideline and corollaries are introduced next.

V. Types of Program Analysis

We have stressed repeatedly that administrators need valid information to make effective decisions, and we have thus far provided three guidelines that need to be followed in all evaluations. Furthermore, we have alluded to the fact that there is a whole field of evaluation that has developed to obtain specific types of evaluation data. We have grouped these techniques into three types of program analysis: process, impact, and benefit–cost. Each type is briefly described below and represents Sections II, III, and IV of the book, respectively.

1. *Process analysis*. Process analysis focuses on program operations. It describes the program and the general environment in which the program operates including who are the persons served, what services are provided, how much does it cost, and how could the program be replicated. You might ask at this point, "why should I read Section II on process analysis?" First, you must describe your program before others will be able to replicate it. Also, process analysis requires systematic data collection and interpretation, efforts that are more involved than the descriptive goals of process analysis at first glimpse imply. A second reason for reading Section II is that sound data-based description is the prerequisite for doing either impact or benefit–cost analyses.

2. *Impact analysis*. Impact analysis focuses on a program's effects or impacts on the targeted population. In its examination of programmatic effects and impacts, this type of analysis also seeks to determine the *difference between the courses of action being considered*. In most cases, this is the difference between the current mix of services and what would happen if a new course of action were undertaken. Your reading of Section III on impact analysis should help you answer questions such as these:

- Did the program have intended effects on outcomes?
- How big are the program's effects?
- How much uncertainty is there surrounding the estimate of each effect?

3. *Benefit–cost analysis*. Benefit–cost analysis focuses on whether a program had impacts that were sufficiently large to justify its costs. All impacts are included here, not just those that are measured or valued in dollars. Benefit–cost analysis can also examine a program's effectiveness and identify alternatives that achieve a given objective at the lowest net cost. As you will see in Section IV, our approach to benefit–cost analysis provides a means for organizing and summarizing information about a program so that it can be used to assess whether a program is likely to be efficient and to determine the nature of its effects on the distribution of income and opportunities. Thus, you might read Section IV to provide you with:

- A convenient summary measure for those impacts that can be measured and valued in dollars.
- A framework for assessing the uncertainty surrounding the various impacts that can be valued.
- A framework for assessing the potential importance of impacts that cannot be valued.

As you read about these three types of analysis, the examples we give, and the guidelines or cautions we present, it is important to remember that analytic techniques cannot make decisions; they can only help smooth and clarify the decision process. In the end, administrators must make decisions that reflect their assessment of the political and social implications of programs as well as of the objective evidence about program impacts. Furthermore, administrators will never have complete information and certainty; thus, decisions must be based on experience and value judgments in addition to evaluation information. The real benefit of formal analysis procedures is for organizing and interpreting available data. Thus, our fourth guideline.

Guideline 4. Program analysis provides rules of evidence for organizing information to make judgments.

VI. Summary

In this first chapter we have suggested that administrators are decision makers and that making decisions requires valid information and evaluations. In many cases, evaluation is informal and quick; but some decisions require more evaluation, since they involve more complex choices, are harder to reverse, or have far-reaching implications. Since administrators are constantly making decisions, and thereby conducting evaluation, we proposed, throughout this and subsequent chapters, to help improve this process by increasing the efficiency of time spent in evaluation, the accuracy of the information employed, and the quality of the decision made.

We organized the material around four areas that both explain our approach to decision making and evaluation in program administration and set the stage for the remainder of the book. These four included the programs, perspectives and the decision environments; potential decision-making dilemmas faced by program administrators; programmatic decisions and evaluators; and our three suggested types of program analysis. As promised, we proposed four guidelines that are involved in all evaluations, and thus they represent rules that need to be followed. These guidelines include:

- As a consumer, the major question you want to pose in reading an evaluation is this: Is there enough certainty and applicability to my situation to act on the evaluator's results?
- As a producer, you want to (1) carefully define the question you want to answer, (2) collect information about that question, and (3) interpret that information, making sure your conclusions follow logically from what you observe.
- The evaluation focus must match the decisions that need to be made.
- Program analysis provides rules of evidence for organizing information to make judgments.

Throughout the chapter we have stressed the dual role that many readers have—that of a consumer and producer of evaluation data. In that regard, we have stressed that most readers will be consumers of all types of evaluations, but will probably not be producers of large-scale evaluations at the demonstration or ongoing stages; rather, most will be producers of small-scale feasibility-stage evaluations. Additionally, we have stressed that regardless of the scale of evaluation, one must use logical thought and follow the scientific method. And finally we have stressed the uncertainty inherent in all we do, and that evaluation is a continuous, dynamic process that flows from one type of analysis to another, depending upon the questions asked and the resources available to answer the questions.

Thus we now move to Chapter 2, wherein we ask you to describe your program's purpose, population, and process, in order to get you started producing program evaluations. In your descriptions, we ask that you be systematic in your collection and analysis of program data, and that your conclusions be based on thoughtful, careful, and honest analysis.

Describing Your Program

I. Overview

The process of evaluation begins with specifying exactly what is being evaluated. This careful specification is necessary in order to make the structured comparisons that lie at the heart of any evaluation. It begins with a statement of the problem or problems being addressed by the program under evaluation; that is, it reviews the program goals. It then outlines the population being served and the services provided. Finally, it considers the process by which the services provided will produce the desired effects on the persons being served.

The mother-in-law test is applicable throughout this specification. The key is be sure that your specification is clear and that it would make sense to a skeptical, but generally reasonable, audience. The description should not rely on unstated assumptions or professional jargon. It should be simple, with the desired results following logically and directly from the nature of the services. The essence of the mother-in-law test is to prepare the specification and then reread it, putting yourself in the position of a person not familiar with your program. Ask yourself, have you set out your goals, is it clear whom you are serving, and have you indicated how the specific services provided will produce the desired results? To help you remember the mother-in-law test, we summarize it for you as follows:

> *Mother-in-law test.* Does your presentation make sense to a person unfamiliar with the day-to-day operation of the program and with program evaluation? Is the presentation straightforward, does it avoid the use of jargon, and does it rely on its own logic rather than relying on unstated assumptions about common practice? Will the presentation convince a willing, careful, but somewhat skeptical reader?

This is a brief, but important chapter. What we want you to do is describe a program's purpose, population, context, and process. We feel this effort is well justified for a number of reasons. First, before you can answer, How am I doing?, you need to explain what you are doing. Second, it may cause you to rethink what you are doing and to consider alternative strategies or techniques. And third, we anticipate that this exercise will show the relevance of the book's contents, even if you get no further in the book. The process of reflecting on your program's goals, the persons it intends to serve (and those it actually serves), the services it provides, and the process by which it is to achieve its goals is a useful evaluation exercise that will improve your program.

Although we return to the area of program description in Chapter 3 and Section II, the above reasons suggest that it is important at this point to go into a room, close the door, and think about your program. In the process, justify what you are currently doing, and then attempt to describe your program to others. The activities suggested in the chapter are similar to what many of you have already done, namely, sat down and described your program to a reporter. We suggest that you use our proposed mother-in-law test, which rests on the assumption that, in describing your program to others, you can expect a somewhat skeptical audience with no special expertise in your program area. We have used this test frequently and find it both reasonable and productive. A corollary is also worth considering: clear writing usually results from clear thinking, and unclear writing always results from unclear thinking.

In the chapter we discuss six factors to consider in describing your program's purpose, population, context, and process. They include (1) the problem(s) addressed, (2) the persons served, (3) the services provided, (4) the evaluation context, (5) the expected outcomes, and (6) the justification for why you think it's going to work. Examples and suggested formats are provided throughout the chapter that will make your work easier. We don't anticipate that you will need much actual data; rather, you will more likely need planning documents, proposals, program records, and interview information. But we do want to emphasize that evaluation techniques are not learned by simply reading books. Rather, they are more like chemistry, in which the lab work—that is, actual hands on experience—is an essential component of the instruction. Thus, readers who ignore the exercises are cheating themselves. We encourage you to consider them, even if you can only sketch out solutions in your mind. Even that little work will help to insure that you understand the concepts being discussed, the advantages of using the techniques outlined in the text, and the problems that will be faced when you try to put the guidelines and procedures into practice.

II. Problem Addressed

As indicated in the previous chapter, human service programs arise to meet the needs of persons in society. The first evaluation step is to identify the specific needs addressed by the program under study. For example, rehabilitation programs are frequently established to improve a person's functional level, labor market behavior, or quality of life; education programs focus on improving academic skills; mental health programs, the problems of adjustment; offender rehabilitation programs, the problems of crime, recidivism, and antisocial behavior; urban development programs, the problems of urban blight, neighborhood preservation, unemployment, and living standards; and senior services, the problems of living arrangements, living standards, and mobility.

As an example of this process, consider the STETS demonstration introduced in the previous chapter. The transitional employment program tested in that demonstration was intended to increase the ability of young adults with mental retardation to obtain and keep jobs in the regular unsubsidized labor market. Specifically, it had four goals:

1. Improve the labor market performance (measured by employment rate, average hours worked, and earnings) of program participants compared with what would have happened in the absence of the STETS services.
2. Reduce the use of alternative service and income-support programs by the participants (again compared with what would be expected in the absence of the STETS services).
3. Increase the extent to which participants lead normal life-styles and have a broader range of choices.
4. Save government and social resources by reallocating funding for services to transitional employment rather than to other more long-term support services, and helping to increase the productivity of the participants.

In addition, there was the implicit assumption that STETS participants would be better off with the increase in employment and independence and the reductions in the use of alternative programs. If these changes failed to increase the well-being of the participants, then the whole rationale for the program would be called into question.

These goals form the basis for the evaluation and so must be carefully specified. As a result it is essential to consider secondary or implicit goals as well as the central goals. For example, the STETS demonstration also had the goal of increasing the body of knowledge pertaining to vocational service delivery to persons with mental retardation. It also hoped to provide seed money for service expansion in the five demonstration communities. The program operators who were funded by the demonstration may have had goals of obtaining general financial support in a time of ever-scarce social service funding or of increasing the prestige of their organizations. Finally, the U.S. Department of Labor (who funded the demonstration) may have wanted to satisfy an interest group that was seeking help; or the Department may have wanted to expand its role into an area more traditionally under the preview of the Department of Education.

The evaluation must sort out through all these direct and indirect goals and select those against which program performance will be measured. The other goals are then used in describing the context of the demonstration because they may effect the extent to which the program achieves its primary objectives. The essence of the process is to consider each group involved in the program (the target population, program operators, funding agencies, and taxpayers in general) and assess why they support the program and what they hope to accomplish. Each of the goals identified in this process should then be assigned a priority and the evaluator must decide which will be included as primary goals and which will be considered in the context.

As a useful exercise, consider a program with which you are familiar and try to specify its goals. Try to be specific in listing the behaviors, opportunities, attitudes, or choices that the program expects to affect. Additionally, think about the public policy or public law that provided the impetus for its creation, and the major problem addressed by the program. To remind you of the importance of the exercise, we have provided space (although it might well be too small for your complete answer) in which to write your response.

Step 1. What is the major problem addressed by the program? Additionally, what behaviors, opportunities, attitudes, or choices does the program expect to affect?

III. Persons Served

A critical element of the program description is to specify the persons to be served. This will provide a key evaluation criterion, namely, is the program actually serving the person it was designed to serve. Knowledge about who is served is also essential for assessing the appropriateness of services and the magnitude of overserved program outcomes. Clearly, the methods used by an employment program, for example, will be viewed differently if that program serves able-bodied welfare recipients or if it serves persons with severe mental or physical disabilities.

Thinking about who is served and who the program was designed to serve is a central aspect of program evaluation and something that can be investigated with relative ease. For example, consider the hospital administrator referred to in Chapter 1, who was interested in the effects of medication reduction, was also interested in the composition of the patient population. In doing a quick record review, he found that on any given day during the last year, nearly 70% of the patients had previously been hospitalized at least once. This information was very valuable in changing the clinical program to focus on more chronic problems. It also changed significantly how he answered the question, Who are the persons served by [your] program?

How you describe the persons served raises a number of questions. You want to give enough detail so that the listener (the reporter or your mother-in-law) understands the participants' major characteristics. A reasonable guideline to suggest is this: What would you want to know about a group of people if you had to design a program that would achieve the goals outlined in Step 1?

One good source of information regarding the persons served is the program's eligibility criteria. In reference to the STETS demonstration, each participant was to meet the following criteria:

- *Age between 18 and 24, inclusive.* This group was chosen to enable the program to focus on young adults who were preparing for, or undergoing, the transition from school to work or other activities.
- *Mental retardation in the moderate, mild, or lower borderline range.* This criterion was indicated by an IQ score of between 40 and 80 or by other verifiable measures of retardation. Specifically because IQ scores have been challenged as a valid measure of employability, special efforts were made to recruit applicants whose IQs were in the lower ranges to ensure that the effectiveness of the program for this group could adequately be tested.
- *No unsubsidized full-time employment of 6 or more months in the 2 years preceding intake, and no unsubsidized employment of more than 10 hours per week at the time of intake into the program.* This criterion was established to limit enrollment to persons who would be likely to need the intensive employment services envisioned in the model.
- *No secondary disability that would make on-the-job training for competitive employment impractical.* While the demonstration was designed to test the effectiveness of the program for a disabled population, it was recognized that some individuals would have secondary disabilities of such severity that these individuals could not be expected to work independently in a regular job. While the projects were required to make such determinations, they were encouraged to apply this standard only in exceptional cases.

Without spending a lot of time digging through files (should you not have easy access to the characteristics of the people served by the program), complete Step 2 as best you can for the program you are describing.

Step 2:   Briefly describe whom the program intends to
serve.

IV. Services Provided

Human service programs are service, or process, oriented; they provide
services to their participants. But don't think about services in a "self-fulfilled
prophecy" sense such that rehabilitation programs rehabilitate, education programs educate, criminal justice programs correct, and so on. Rather, think about
specific participant-referenced activities performed usually by program staff on
behalf of the participants. For example, most human service programs with
which the authors are familiar offer some form of the following participant-
referenced services: assessment or evaluation; education, training, or therapy;
living or employment environments; case management support or assistance;
transportation; and follow-up.

Programs, whether they have a singular or multiple purpose, need to define
clearly the services they provide. Let's return to the STETS example. Here, the
services provided to participants consisted of three sequential phases.

- *Phase 1* involved assessment and work-readiness training. This phase
  combined training and support services in a low-stress environment, the
  goal of which was to help participants begin to develop the basic work
  habits, skills, and attitudes necessary for placement into more demanding
  work settings. This preliminary stage, which was limited to 500 hours of
  paid employment, occurred in either a sheltered or nonsheltered work
  setting, but in all cases, the participants wages were paid by the project.
- *Phase 2* involved a period of on-the-job training in local firms and agencies. During this stage, participants were placed in nonsheltered positions
  that required at least 30 hours of work per week, and in which, over time,
  the levels of stress and responsibility were to approach those found in
  competitive jobs. Wages were paid by either the project or the employers
  or some combination of the two. The STETS program provided workers
  in Phase 2 with counseling and other support services, and it helped the
  line supervisors at the host company conduct the training and necessary
  monitoring activities.
- *Phase 3,* which included postplacement support services, began after
  participants had completed the Phase 2 training and were performing their
  job independently. The purpose of this phase of program services was to
  insure an orderly transition to work by tracking the progress of participants, by providing up to 6 months of postplacement support services

and, if necessary, by developing linkages with other local service agencies.

Section II goes into considerable detail in describing, quantifying, and costing out specific services provided to participants. But for the time being, simply list those services your program provides to its participants.

Step 3. List the specific services provided to the program's participants.

V. Evaluation Context

Social programs do not operate in a vacuum. They interact with other programs and the general social and economic conditions of the communities in which they operate. Such interactions will affect program performance and the generalizability of the findings for particular projects. Thus it is essential to document the context in which the program or policy under evaluation was implemented.

The STETS demonstration was fielded in 5 large cities and ran from the fall of 1981 through December, 1983. During this period the economy went through a severe economic recession, potentially making it difficult to find jobs for the program participants. In addition, each of the 5 cities in which the demonstration projects operated had its own pattern of services for assisting persons with disabilities. The projects could work with the other programs in their community to supplement the demonstration resources to varying degrees and could count on differing degrees of community support for the demonstration goals. Both the economic conditions and the other community programs needed to be documented because they affected the costs and impacts of the STETS services.

Administrators seeking to use the STETS results must have access to this information if they are to assess the degree to which the STETS experience is applicable to their own programs. Similarly, an administrator making a decision about moving residents of an institution for persons with severe mental illness into the community would want to know the context of previous deinstitutionalization efforts. In particular, if he or she found a study in which such moves went smoothly, it would be critical to know if the residents who moved into the community could rely on a network of community mental health centers or if the institution had to provide such supports. Such environmental support could make a major difference on the effectiveness of the deinstitutionalization program and its costs.

Another aspect of the program context pertains to the historical antecedents of the program under study and the experiences of other programs similar to that program. Each human services program develops in a particular time frame and in response to specific needs and social policies. For example, the STETS demonstration was undertaken to determine the effectiveness of transitional employment programs in integrating mentally retarded young adults into the economic and social mainstream. Interest in these transitional employment programs arose from public policy and public laws (such as The Rehabilitation Act of 1973 and the Education of All Handicapped Children Act of 1975) that recognized the rights and abilities of disabled persons who had traditionally been underemployed. The demonstration also drew on theory and previous experience suggesting that:

- Behaviorally based training strategies were an effective training tool.
- Transitional employment programs were effective with other disadvantaged populations such as long-term welfare recipients, school dropouts, exaddicts and unemployed workers.
- Transitional employment pilot programs for young mentally retarded adults were both feasible and promising.

There are a number of reasons why we feel it is essential to understand a program's context. One is to evaluate fully and correctly its intended purposes and anticipated outcomes. Second, generalization from its results is dependent upon generalizing to similar contexts. And third, there are a number of pitfalls if one does not keep the program's context in mind. For example, deinstitutionalization occurring in an area with good mental health services will probably have different results than if it occurs in an area where mental health services are poor and minimal. The time frame is also important. Work for the retarded in the 1960s meant sheltered workshops; today, work means a paid job in the regular work force. Thus, one frequently needs to look back 10–15 years to understand fully what the program's goals and program personnel's perceptions were at that time.

Thus, the evaluation should contain a brief discussion of the historical context within which the respective program was developed and the current context within which it is evaluated. It's time now to think about the program that you are describing, and more specifically, to think about its historical and present contexts. Specific factors to consider include (1) when did the program develop, (2) what were the public policy issues around which the program developed, (3) what theory and previous experiences were drawn upon in developing the program, and (4) what aspects of the economic and programmatic environment affected the program under study?

Step 4. Briefly describe the economic and programmatic environment within which the program must operate.

VI. *Expected Outcomes*

In thinking about and describing the program's expected outcomes, it is important to link outcomes to goals. For example, if one of the program's goals is to increase employment, then the outcomes of interest are placement into jobs, hours of work, wages, and average weekly earnings from these jobs. Thus, one's programmatic outcomes flow from the program's goals. However, since human service programs tend to be process oriented, many programs have difficulty thinking about or describing their anticipated or expected outcomes. If you are having this trouble, we offer Table 2.1. If you are an educator, you will probably focus on academic measures such as skills, grades, and standardized test scores; mental health programs may focus on functional level, recidivism, or symptom reduction; rehabilitation programs might measure labor market behaviors; criminal justice programs, recidivism and labor market behavior; programs for the elderly, activities in daily living or institutionalization rates; and programs for the severely or profoundly handicapped, functional level, employment, community interaction, and general quality of life.

Before asking you to list the expected outcomes from the program you are considering, we would like to present an example along with two important points to consider in measuring expected outcomes. The example relates to the STETS demonstration that analyzed outcome data in four major areas that reflected its goals (which were presented earlier): labor market behavior; training and schooling; public transfer use; and economic status, independence, and lifestyle. Labor market behavior was measured by the percentage of persons employed in a regular job and average weekly earnings from the job. The analysis of impacts on training and schooling examined whether the participants used other training programs or schooling. Changes in receipt of public transfer were measured using data about use and monthly income from Supplemental Security Income (SSI), Social Security Disability Insurance (SSDI), Aid to Families with Dependent Children (AFDC), or other cash transfer benefits. Economic status was evaluated in terms of total personal incomes of participants, whereas independence and life-style were measured by financial management activities (pays for purchases, banks or pays bills by self), and living arrangements, respectively.

The two important points to consider about expected outcomes include measurement error and the distinction between outcomes and impacts. As dis-

Table 2.1. **Potential Participant-Referenced Outcome Measures for Different Program Types**

Vocational rehabilitation	Education	Mental health	Chemical dependency	Corrections	Housing assistance	Senior services
Adaptive behavior level changes	Attendance	Functional level	Functional level	Recidivism	Living standards	Activities of daily living
Quality of life indicators	Grades	Symptom reduction	Symptom reduction	Diversion programs	Family development	Living arrangements
Labor market behavior	Adjustment ratings	Labor market behavior	Recidivism	Criminal offense record	Integration	Global life satisfaction
Living arrangements	Credits	Target complaints	Labor market	Prosocial behavior	Housing and living conditions	Environmental hazards
Community integration activities	Diploma	Current and past psychopathology	Self-concept	Life domain changes		Mortality/longevity
Reductions in public transfer monies			Motivation	Home		Institutionalization
				School		Health status
				Work		
				Employment		

cussed in the previous chapter, there is error inherent in all measurement, and this error creates uncertainty. Additionally, the measures selected may imperfectly reflect the program's goals, and this should also be taken into account. And finally, outcome measures need to be quantified and defined clearly. Unclear or ambiguous definitions will create problems in measuring outcomes and in interpreting what they mean. Such fuzziness, which is an available source of uncertainty in evaluations, can be eliminated with careful thought and attention to detail. It does not require massive resources, only clear thinking.

The second point relates to distinguishing between outcomes and impacts. *Outcomes* pertain to observations of individuals' activities over time, whereas *impacts* refer to how those outcomes differ from what would have happened if the program (or service) was not there. The distinction between outcomes and impacts is shown in Figure 2.1.

Placement rates, for example, measure an outcome; if your program improves those rates over what they would have been without your program, they represent an impact. Therefore, one must evaluate changed employment rates relative to the comparison situation. Sometimes, strange things happen when evaluating a program's expected outcomes. The program may well report good placement rates, but the persons served may have been able to find jobs even in the absence of the program. Thus, the program's impact may have been less than that suggested by the outcomes for program participants. Analogously, placement rates may not reflect the program's goal. For example, if the program's goal is long-term employment, then short-term placement rates may not be good enough to evaluate the impact of long-term employment. We return to the distinction between outcomes and impacts in Chapter 3 and especially in Section III on impact analysis. For the time being, we simply want to sensitize you to the distinction and stress that you focus only on outcomes as you complete Step 5. In completing the step, think about expected programmatic outcomes; describe them concisely and quantitatively; and then list them.

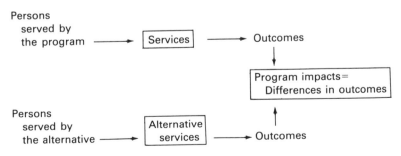

Figure 2.1. Distinguishing between outcomes and impacts.

Step 5.   List the program's expected outcomes.

VII. Justification (Rationale)

The final component of the description is the model or rationale underlying the other components. It is a statement about why you think that providing the stated services to the target population should produce the desired outcomes and impacts. Without this underlying model, the other components of the description are only a wish list.

The model need not be complex or arcane. It can be based on a theory or on evidence from other programs. For example, economic theory argues that one way to curtail crime is to increase the relative attractiveness of legitimate work opportunities. An offender rehabilitation program model based on this theory might increase the attractiveness of work by offering a potential criminal better employment opportunities through placement assistance, training, or education. Conversely, it might lower the attractiveness of crime by increasing the probability of apprehension or the penalties associated with conviction.

Alternatively, a model could be based on observations of pilot programs or the behavior of groups of individuals. For example, the offender rehabilitation program might offer employment opportunities to persons released from prison because previously such programs reported finding that such employment reduces recidivism.

In any event, the key is to have a rationale for why the program under study should work. This rationale will help define the appropriate services for the target population and will also be useful for interpreting the evaluation results.

And now comes the hard part, for you need to justify why you think the program is going to work. In providing this justification, you can draw upon three sources: theory, program evaluation studies in the literature, and previous program history. The basic question to focus on is why you think that the services provided to persons in the program will alleviate the problem addressed. The process is shown diagrammatically in Figure 5.1 (p. 102).

This is where your role as a consumer of program evaluation studies becomes very important. Additionally, it might be interesting to reread the original program design documents or funding proposal to review how the proposed program was justified. In any event, let's look at a few examples before asking you to provide the justification.

In reference to STETS, evidence for linking problem, persons, services, and expected outcomes came primarily from published feasibility studies that showed the potential value of transitional employment programs for enabling mentally retarded persons to become more self-sufficient. Published efforts included university-based programs, programs operated by foundations and private organizations, and national programs such as The National Supported Work Demonstration and the Comprehensive Employment and Training Act. This program history provided evidence that transitional employment programs were feasible and could be used to justify why program administrators thought that the specific services provided to the specific participants should produce the anticipated outcomes and alleviate the problem being addressed.

In justifying why you think the program you're considering is going to work, you might discover some interesting facts. For example, think about a teenage unemployment policy that is directed at reducing unemployment but doesn't provide adequate job-training opportunities. In thinking logically about justifying the connection between problem, services, and expected outcomes, an administrator of such a program may well find that the program is geared primarily toward keeping people off the street rather than improving long-term employment. In this case, it should be no surprise if the program does not change the long-term employment rates for program participants. Similarly, an administrator whose program goal is to provide job placement and increase economic self-sufficiency to handicapped persons may find that the services provided are make-work activities rather than job training; hence he finds himself essentially providing baby-sitting services to produce these expected outcomes. And what about the mental hospital administrator who has just discovered that 70% of his patients are repeats? If his perceived goal was permanent deinstitutionalization, then his services and outcomes are not addressing that public policy issue. This program must rethink its goals and either change methods or adopt different goals.

Thus, we come to the sixth and last step in describing the program's purpose, population, context, and process. And that is to justify why you think it's going to work.

Step 6.   Justify why you think the program is going to work.

VIII. Summary

We anticipate that the six-step process you have just completed will be beneficial to you, regardless of whether you are an evaluation producer or consumer. As you described the program's purpose, population, context, and process, you may have seen flaws in the ability to relate goals to services to outcomes. Every now and then you may have seen that a series of incremental changes took you where you didn't want to be. Hence, this six-step analysis was probably very beneficial. It may even have resulted in programmatic changes. If the process has been beneficial and the program improved as a result, you may not have to go any further in your evaluation.

In summarizing the chapter's contents, we would propose that you use a format such as suggested in Survey 2.1. The survey requires that you summarize the previously written material from each step into the matrix provided. We have given the survey a number because we refer back to it frequently throughout the remainder of the book.

Survey 2.1. **Steps Involved in Describing Your Program's Purpose, Population, and Process**

Step and activity	Your answer
1. What is the major problem addressed by the program?	
2. Briefly describe whom the program serves.	
3. List the specific services provided to the program's participants.	
4. Briefly describe the program's context.	
5. List the program's expected outcomes.	
6. Justify why you think the program is going to work.	

We hope you have found the chapter interactive and enjoyable. It's the beginning of using this book as a field guide. We have suggested that it is very worthwhile for you to sit down, think about and justify what you are currently doing, and then attempt to describe your program to others. In that description, we suggested that you use our proposed mother-in-law test, or, for that matter, describe your program and its rationale to your mother-in-law.

3

Guidelines Governing Program Evaluation

I. Overview

Program evaluation involves the systematic collection and analysis of information about alternatives. Evaluations may be informal and quick, or they may be complex, highly structured efforts. But regardless of their size, all evaluations are subject to the same general rules. Even though an evaluation may have a small budget or a short time frame, evaluators need to follow the same rules or guidelines so as to provide accurate and precise answers to the comparisons they are trying to make.

In this chapter we review those guidelines that pertain to all evaluations, regardless of their size or complexity. Thus, this chapter foreshadows what you will read in detail later in the book. Although we realize that small-budgeted evaluations must accept more uncertainty, they should not employ watered-down rules or guidelines. There are a number of reasons why you should read this chapter. First, these are important guidelines for you to use in judging either someone else's or your own evaluation. Second, by being familiar with the guidelines, you will be better able to allocate your evaluation resources. And third, if you do need to make choices in the evaluation, the guidelines will suggest what the effects might be. For example, you may not be able to marshall considerable evidence, which will affect how you interpret the findings; but it will not affect what we refer to as the "setup," which is what you just described in Survey 2.1.

You should also read the chapter to strengthen your role as an evaluation producer or consumer. Chapters 1 and 2 introduced you to these roles. We suggested that as a producer, you will most likely be involved in feasibility-stage evaluations that attempt to show the feasibility or potential of your program. If you do attempt this, however, expect a skeptical audience that will want you to justify or support your contention that your program produced the desired effects. Therefore, in this chapter we discuss a number of guidelines governing program evaluation that will help you in your role as a producer. As a consumer, these rules will also help you evaluate others' contentions that their programs produced the desired effects.

Program evaluation is primarily a structured comparison. Thus, as we proposed in Chapter 2, the first step of an evaluation is to specify the program or policy being evaluated and the program or option with which it will be compared. This specification should include information on such factors as the persons being served, the treatments being offered, and the environment in which the program or policy will operate. These two alternatives—the program and the comparison situation—define the scope and ultimately the results of the evaluation. All components of the evaluation, including interpretation of the findings, must be undertaken in relation to these two alternatives.

To assist you in developing and evaluating these alternatives, we discuss a number of guidelines, listed in Table 3.1. They are included in the setup, marshalling the evidence, and interpreting the findings. The setup, which was discussed in detail in Chapter 2, includes the problem addressed, persons served, services provided, evaluation context, expected outcomes, and justification or rationale that links the intervention or services to the expected outcomes. The major purpose of the setup is to define the structured comparisons, pose the analytical questions, and rationalize the expected outcomes from the services provided. The second set of guidelines relates to collecting, or marshalling, the evidence to justify or support the contention that the program described produced the desired effects. Once the evidence is marshalled, it is interpreted within the context of the structured comparisons, the program's environment, and any limitations or weaknesses of the evaluation performed.

We suggest that the guidelines listed in Table 3.1 are applicable to any kind of program analysis, including process, impact, or benefit–cost. Further, we suggest that they apply regardless of the level of intensity or precision. Persons asking very simple process questions such as, who is served by this program? should follow the rules just as closely as those asking the question, how big are the impacts (or benefits and costs) of this program? And finally, we offer a caution regarding these rules: they are tightly interrelated, since the setup drives marshalling the data, which in turn results in the findings that must be interpreted.

In reading the chapter, we hope you keep this thought in mind: you won't have to develop the transistor every time you go into the laboratory; that is, policy evaluation and development, like science, is incremental. Evaluation efforts often pay off in helping to make small decisions. Or many small evaluations may be needed to produce a body of evidence sufficient to bring about policy change. Regardless of the size of your evaluation effort, we ask only that you use logical thought and pay attention to the details of your evaluation so that you can tell your readers or colleagues that you have attended to the technical criteria and basic principles governing program evaluation.

Table 3.1. **Guidelines Governing Program Evaluation**

The setup	Marshalling the evidence	Interpreting the findings
1. Problem addressed	1. Define outcomes.	1. Have correspondence between the setup and marshalling the evidence.
2. Persons served	2. Select measures that reflect initial intent.	
3. Services provided	3. Describe geographical context of program.	2. Attribute changes in outcome to the intervention.
4. Evaluation context	4. Measure services delivered and duration of services/interventions.	3. Interpret conclusions within the program's context.
5. Expected outcomes	5. Measure changes in outcomes.	4. Address the issue of uncertainty.
6. Justification (rationale)	6. Collect data on both sides of the structured comparison.	

II. The Guidelines

A. The Setup

In the setup, you need to attend to those factors you were introduced to in Chapter 2 and that you described in Survey 2.1. Because of their crucial importance in any evaluation effort, we want to return to them briefly in this chapter to indicate both how they relate to the guidelines summarized in Table 3.1 and how they provide the basis for conducting or interpreting program evaluation. In the setup, you need to include those six factors listed in Table 3.1: (1) the problem addressed, (2) the persons served, (3) the services provided, (4) the evaluation context, (5) the expected outcomes, and (6) the justification or rationale.

1. The Problem Addressed and the Goal of the Evaluation

The evaluation's goal or purpose includes a well-defined statement of what you are trying to show with the evaluation, what you are trying to compare, and what you are trying to test with the data. Let's return to the STETS and National Long-Term Care demonstrations to present and discuss examples of each of these requirements.

At the outset, the goal or purpose of the evaluation should be stated clearly. The STETS demonstration project was: designed to test the effectiveness of transitional-employment programs in enhancing the economic and social independence of mentally retarded young adults.

Similarly, the National Long-Term Care demonstration program was designed to test whether a managed approach to providing community-based long-term care could help control costs while maintaining or improving the well-being of its clients and their informal caregivers.

The evaluation's purpose sets the stage for everything that follows, and thus must be clearly stated and understood, whether you are a producer or consumer. Thus, if you are a producer, be sure that you state your evaluation's purpose clearly and measurably; as a consumer, carefully check the introductory section of the article or report to find the evaluation's purpose or goal statement.

The second requirement relates to what the evaluation is trying to compare. It was stated previously that evaluations are structured comparisons. We make implicit comparisons all the time such as where to shop, whom to hire, or whether to add an additional support service. In program evaluation, however, we are expected to make formal comparisons between alternative courses of action, be they treatments, interventions, or policies. Thus, we need to be very clear about what the treatment, intervention, or policy is and with what we are comparing it.

The STETS project, for example, compared what happened when mentally retarded young adults received STETS services with what would have happened in the status quo—that is, in the absence of the STETS services. In this type of comparison situation, the persons who enrolled in STETS might have used other services and programs or might have received no special services. In many cases, this is the nature of the comparison implied by program decisions: should a new service or program be added to the existing system, or should things be left as they are?

In the Long-Term Care project, the comparison involved the following:

- *The basic case management model,* which augmented the case management intervention with a small amount of direct service purchasing power to fill service jobs, versus the status quo.
- *The financial control model,* which through the pooling of categorical program funds permitted channeling case managers to order the amount, duration, and scope of services that they deemed necessary, versus the status quo.
- *Impacts* under basic model versus impacts under financial control model.

A comparison of one model against the other would have been inadequate, since it would not have indicated whether either was better than the current system.

The goal of the evaluation should also include a general statement regarding what you are trying to test with the data. This statement should not include specific statistical tests that will be used to analyze the data; rather, it should be a general statement, such as the following made in reference to the STETS and Long-Term Care projects:

> *STETS.* This demonstration will greatly expand our knowledge about the implementation of transitional employment for this target population, and will document the effectiveness of such programs to increase employment, earnings, and independence as a means of increasing well-being. (Kerachsky *et al.,* 1985, p. 3)

> *Long-Term Care.* It is hoped that both models (basic case management and financial control) will enable impaired elderly persons to remain in the community rather than enter a nursing home. In that regard, the demonstration tried to reduce institutionalization as a means of increasing well-being and reducing overall LTC expenditures. (Thorton & Dunston, 1986, p. 4)

Thus, the goal of the evaluation, as one component of the setup, should provide a clear indication of the evaluations's purpose, structured comparisons, and a general statement of what is being tested with the data. These requirements constitute our fifth guideline:

Guideline 5. The goal statement of your program evaluation should include (1) a well-defined purpose, (2) clearly defined structured comparisons, and (3) what you are trying to test with the data.

---

2. The Persons Served

We discussed in Chapter 2 how one might describe the persons a program serves, and we suggested including the program's eligibility requirements as one good way to do this. Frequently, however, one will also need to provide more specific details about those persons. As an example of how this might be done, we offer the following sample descriptions from the STETS and Long-Term Care projects.

STETS was implemented in five U.S. cities. The final research sample consisted of 437 individuals—226 experimentals and 211 control group members. Of this total, 59% of the evaluation sample were male, and half were from minority ethnic or racial groups. Sixty percent had IQ scores that indicated mild retardation, and 12% had IQ scores indicating moderate retardation. About 80% lived with their parents, and another 10% lived in supervised settings. Less than 30% were able to manage their own finances. Nearly two-thirds were receiving some form of public assistance; one-third were receiving either Supplemental Security Income (SSI) or Social Security Disability Insurance (SSDI). Prior vocational experience was limited primarily to workshop and activity centers, and about one-third had had no type of work experience in the 2 years prior to enrollment. However, this group was interested in making the transition to unsubsidized employment, as evidenced by the fact that 70% had worked or participated in an education or training program in the 6 months prior to applying to STETS.

The Long-Term Care project involved approximately 5000 persons who were in either the basic case management or financial control model group. The basic model group averaged 79 years of age at baseline; 56% had very severe or extreme impairments in the measured activities of daily living (ADL) including eating, transfer, toileting, dressing, and bathing. Forty-three percent of the clients reported more than three unmet needs at baseline, and 13% were dissatisfied with their service arrangements. The average monthly income at baseline was $528, and 60% reported being "pretty or completely satisfied" with life. In comparison, clients in the financial control model averaged 80 years at baseline, and 60% had very severe or extremely severe ADL impairments. Fifty-three percent reported more than three unmet needs at baseline, and 11% were dissatisfied with their service arrangements. Their monthly income at baseline was $547, and 52% reported being "pretty or completely satisfied" with life.

The above two paragraphs should provide some guidance as to the type of client descriptions that are appropriate in describing whom your program serves. This information is essential, as we see in subsequent chapters, in determining whether the participants in the program being evaluated are similar to those with whom you are working. You must be careful not to compare apples and oranges.

3. Services Provided

Human service programs generally pride themselves on the services or interventions they provide. Thus, the program evaluation should include a detailed description of these interventions or services. Table 3.2 summarizes the interventions employed by the STETS and Long-Term Care projects. The STETS program experience for clients consisted of these sequential phases: (1) assessment and work-readiness training, (2) transitional jobs, and (3) follow-up support services. The Long-Term Care project involved the concept of channeling, which was designed to affect client well-being, service use, and the cost of care primarily by coordinating the long-term care service needs of clients with the services available in the community. In developing the demonstration, the seven core channeling functions listed in Table 3.2 were identified as the minimum set of functions deemed necessary to achieve this objective.

4. The Context of the Evaluation

Social programs develop out of human needs and within social, political, and historical contexts. Examples include social security, antidrug programs,

Table 3.2. **Examples of Interventions or Services Provided: The STETS and Long-Term Care Demonstration Projects**

STETS program phases	Long-Term Care project core functions

1. Assessment and work-readiness training (service lasts up to 500 hours of work)
  a. Combined training and support services to help participants develop work habits and skills
  b. Participants engaged in at least 20 hours of productive work weekly
2. Transitional jobs (must be completed within 12 months of enrollment)
  a. Provided at least 30 hours per week of on-the-job training and paid work
  b. Participants were provided counseling and other support services
3. Postplacement support services
  a. Provided up to 6 months of postplacement support services
  b. Developed linkages with other local service agencies if necessary
	1. Casefinding/outreach—identified and attracted the target population
	2. Screening—determined whether an applicant was part of the target population
	3. Comprehensive needs assessment—determined individual problems, resources, and service needs
	4. Care planning—specified the types and amount of care to be provided to meet the individual needs of individuals
	5. Service arrangement—implemented the care plan through both formal and informal providers
	6. Monitoring—ensured that services were provided as called for in the care plan or were modified as necessary
	7. Reassessment—adjusted care plans to changing needs

community-based mental health and mental retardation programs, special education or rehabilitation programs, area agency on aging programs, antipoverty and job programs, crime prevention programs, and many more. Each of these program types has developed and changed over time and in response to specific needs and social policies. One needs to understand these needs and policies to fully evaluate the intended purposes and anticipated outcomes from the program. Therefore, the evaluation should contain a brief discussion of the historical context within which the respective program developed and the current context within which it is evaluated.

In reference to STETS, transitional employment programs for mentally retarded persons have a relatively brief history, dating back only to the 1970s. Several factors related to the historical and present contexts were especially influential in the STETS evaluation study. First, attitudes have changed considerably over the last 15 years regarding the rights and abilities of mentally retarded and other handicapped persons to participate more fully in society and to make substantial contributions to their own support. Among the prominent evidence of this shift is the Rehabilitation Act of 1973; provisions in the Vocational Education Act, the Comprehensive Employment and Training Act, and the Job Training Partnership Act that encouraged the participation of handicapped individuals in education and training programs; the Education of All Handicapped Children Act of 1975; and the 1980 and 1984 amendments to the Social Security Act, whose purposes were to reduce the work disincentives created by the SSDI and SSI programs.

A second relevant factor was that, despite these federal efforts, a small portion of mentally retarded young adults were employed in regular, unsubsidized jobs. These persistently high unemployment rates, together with the substantial federal outlays for income support and special education services to mentally retarded persons, have fostered a growing emphasis on intervention strategies, including transitional and supported employment.

A third factor that fostered the demonstration was that two independent bodies of evidence suggested that transitional employment was a potentially effective way to facilitate the transition of many mentally retarded young adults from school or workshop/activity centers into regular competitive employment. First, the results of the National Supported Work demonstration showed quite clearly that transitional employment programs could be effective in mitigating the employment problems of other seriously disadvantaged subgroups, and that the effectiveness of the programs tended to be greater among the more disadvantaged subgroups of the target populations served (Hollister, Kemper, & Maynard, 1984). Second, a number of relatively small transitional employment programs (Hill & Wehman, 1985; Rusch & Mithaug, 1980; Vera Institute of Justice, 1983) for mentally retarded adults, many of whom were young, have demonstrated the operational success of such efforts.

Similarly, the Long-Term Care project arose in a period of growing concern about the quality of life of elderly persons and the long-term costs incurred when elderly persons are placed out of their homes into nursing homes. The interested reader is referred to the Long-Term Care report (Kemper, 1986) for a detailed discussion of the evaluation's context such as that provided above for the STETS evaluation. The point we wish to stress is that one's program evaluation should provide a historical perspective of the program and why the evaluation is being done now. The material can come from published literature or from other sources.

There are a number of pitfalls if one does *not* keep the program's context in mind. Four examples are discussed briefly below:

1. Impacts may be affected by context and therefore need to be interpreted with the context in mind. For example, STETS was fielded during the recession of the early 1980s. Thus, it tried to improve employment and earnings during a time when unemployment and competition for jobs were high. Had it been tested in better economic conditions, results *might* have been better (this is an aspect of uncertainty).
2. Programs fielded in cities that rely on an extensive transportation system or other urban infrastructure may not be straightforwardly transplanted to the suburbs or rural areas.
3. In different time frames, there may be slightly different meanings to words. For example, in 1960, "vocational activity" meant a workshop; in 1980, it means work in the regular labor market. The changing meanings of words is important to consider when viewing the results of previous evaluation efforts.
4. The context may imply something about the population. A program fielded in one area may not have the same results when replicated in a different area. This is clearest when trying to interpret results from other countries. Also, there have been programs fielded in areas where the population may share a different cultural orientation than other areas. A hispanic community may differ from a black, Chinese, or Korean neighborhood.

It is in these kinds of concerns—that there are special features of the context in which the program under study was fielded—that may limit the generalizability of the evaluation findings. The importance to a program's evaluation is reflected in our sixth guideline:

Guideline 6. Describe the context, particularly those special aspects that might limit generalizability, within which the evaluation is to be interpreted.

---

5. Expected Outcomes

In thinking about the program's expected outcomes, let's briefly review the distinction between outcomes and impacts as discussed in Chapter 2. Outcomes are the activities, attitudes, or behaviors the program expects to affect, whereas impacts refer to how those outcomes differ from what would have happened under the comparison situation (see Figure 2.1). At this point in your evaluation, think about outcomes. You may want to refer back to Survey 2.1 to see what you said about the program's anticipated outcomes.

Outcomes of interest in the STETS project, for example, included employment, dependency on welfare, self-sufficiency, and involvement with non-handicapped persons. Analogously, the anticipated outcomes from the Long-Term Care project included unmet needs, system expenses, the quality of life, confidence in the system, and extended mortality.

In stating your expected impacts, it is important to rationalize them on the basis of some kind of a model. Let's look at two examples from the STETS evaluation. Increased employment in unsubsidized, competitive jobs was the primary objective of STETS. The demonstration was based on the hypothesis that STETS would enhance the ability of participants to obtain and hold jobs in the regular labor market. It was also hypothesized that STETS should have an impact on earnings and hours, since the training and work experience would allow participants to work more hours and earn more money. Wage rates were assumed to be affected, since individuals should be able to perform more capably in jobs, thereby earning higher wage rates. However, some marginal workers (individuals who would not be able to work in competitive jobs without STETS services) were also expected to be able to obtain jobs subsequent to their STETS experience. Therefore, for experimentals as a group, the higher wages of more-able individuals, who performed more capably in jobs, would tend to be offset by the lower wages associated with some less-able individuals, obtaining jobs for the first time.

The second STETS example relates to expected training and schooling outcomes. By definition, STETS participation should produce a short-term increase in the use of training programs, since it is a training program. However, the use of training programs other than STETS should decrease. In the long run (beyond the period of STETS eligibility), STETS was hypothesized to reduce the use of all training programs as the result of increased employment. The effects on school attendance were expected to be consistent over time; participants were hypothesized to be less likely to attend school, both during and after program participation.

We cannot stress enough the fact that there is no substitute for looking at details and working through very carefully what your expected impacts are. Similarly, the expected impacts may not be what you expect on first glance. For example, the program may be expected to get people jobs, but the wage rate may go down because you have brought more persons into the labor market. Thus, it is also important to indicate the sample you are considering. Bringing low-wage earners into the job market may increase placement rate, but not necessarily average wages.

The anticipated impacts are usually expressed in the form of expectations that are tested statistically and summarized under marshalling the evidence. For example, it was expected that the training and work experience of STETS would allow participants to work more hours and earn more money. We develop the notion of research hypotheses in more detail in the section titled Impact Analysis; for the time being, let's focus merely on the relationship between the expected impacts and the rationale linking the intervention or services to the impacts.

6. Justification (Rationale)

One's task with justification, which represents the last component of the setup, is to rationalize or explain how the services provided to these participants should result in the expected outcomes. The two primary sources for constructing this rationale are published evaluation results (evidence) and theory. To explain how services provided resulted in expected outcomes, STETS, for example, used theory as well as published evaluations of the National Supported Work demonstration and a number of relatively small transitional employment programs for mentally retarded persons that had demonstrated the operational success of transitional employment programs. Similarly, the Long-Term Care project used previous evidence and logical reasoning to link the interventions to expected outcomes. The basic case management model tested the premise that the major problems in the current long-term care system pertain to insufficient information, access, and coordination. It was reasoned that these problems could be largely solved by client-centered case management. Similarly, it was reasoned that the financial control model would represent a more fundamental change in the current long-term care system by broadening the range of available community-based services.

7. Conclusion

In summary, the setup is a large and important component of a program evaluation. The component includes the problem addressed, persons served, services provided, evaluation context, expected outcomes, and the rationale linking the intervention or services to the outcomes. The setup is the foundation for the rest of the evaluation, and we refer to it throughout the book. Because of its importance, we offer the following guideline:

> *Guideline 7.* Your setup should include the problem addressed, the persons served, the services provided, the evaluation context, the expected outcomes, and a rationale linking the intervention or services to the outcomes.

Once the setup is complete, one then needs to either provide or look for the evidence that supports the setup and that justifies the contention that the intervention produced the desired effects. We refer to this justification or support process as marshalling the evidence. In moving from the setup to marshalling the evidence, we would like to reformulate those aspects of the setup discussed in this section into a format wherein we focus on the interrogatories of the evaluation. This conversion facilitates our discussion of the two remaining guideline areas— marshalling the evidence and interpreting the findings. The interrogatories for the setup are summarized in the left portion of Table 3.3.

Table 3.3. **Interrogatories concerning the Setup and Marshalling the Evidence**

Interrogatory	The setup	Marshalling the evidence
Who	Persons served	Can you describe who was served by the program under the study? What are their characteristics and how were they selected to be in the program?
What	Intervention/services received	Can you describe the intervention provided to these persons, including its intensity and duration?
	Outcomes from services/status quo	Can you describe the outcomes that resulted from the provision of services and the outcomes that would have occurred if the services were unavailable?
Where/when	Evaluation context	Can you describe the context in which the intervention was fielded?
How/why	Details of the intervention/services provided	Can you describe how the intervention took place and why it had the effects you claim it did?
	Rationale linking services to outcomes	

B. *Marshalling the Evidence*

Marshalling the evidence involves collecting data to support the objectives set out in the setup. This is an active process that in part involves the following:

1. You need to support the setup; namely, do these services provided to these individuals produce these *changes* in outcomes (compared to the comparison situation)?
2. Review the setup to see what data you need. For example, if the goal of the program is employment, then data should be collected on participant employment. If the goal is to reduce institutionalization, then measure that.
3. Remember that impacts will be affected by many factors, so you need data on them as well. Examples include participant characteristics and contextual variables such as geographical area and time period.
4. The same data should be collected for both the treatment and comparison situations. This is frequently the hardest step and involves getting information on what persons would have done without the program. Fortunately, other sources of information are possible for the comparison situation, and we return to this issue in Chapter 8.

How you go about marshalling the evidence depends largely upon whether you are a consumer or producer of the evaluation. As a consumer, you will use the same setup, because you need to identify and specify your decision and comparison situations. The consumer marshalls evidence by checking to see that everything is in the article or report. In regard to interpretation, the consumer worries about applicability of the evaluation findings to the consumer's decision.

As a producer, marshalling the evidence is an *active* process that involves pounding the pavement and getting the data. This is an action task rather than a thought task; it is where the evaluation will use up resources. Although we present appropriate data collection procedures in Sections II–IV, it is important at this point for the producer (and the consumer who is a producer on a small scale) to know that there is a link between the setup and the interpretation—and that link is marshalling the evidence. It is also important to realize that it is this stage that builds the case for or against a program. One rule for a producer is to include the information you would need if you were a consumer; that is, what would you want to know before you put your scarce resources into a program.

We would like to use an analogy to help you remember the important aspects of marshalling the evidence. The analogy is based on the following quote by Jules Henri Poincaré (1968):

Science is built up with facts, as a house is with stones. But a collection of facts is no more a science than a heap of stones is a house. (p. 80)

In our approach, the setup is the blueprint for determining the types of stones that will be needed and assembling them. The stones themselves are collected in the marshalling of the evidence. The interpretation is the process of putting everything together and assessing the overall strength of the building. We follow this analogy throughout this section on marshalling the evidence.

One way to ensure that all of the necessary stones are collected is to follow the interrogatories listed in Table 3.3 and see that each one is addressed. Thus, the interrogatories help the consumer and producer ensure that all bases are covered. In summary, if you are a producer, you will need to have the blueprints, collect the stones, and build the house; if you are a consumer, you also will need to have the blueprints, but you will need to be a building inspector to ensure that all the stones are there, the blueprint was followed, and the house is livable.

1. Who

The questions you need to answer under "who" include: (1) can you describe who was served by the program under study, (2) what are their characteristics, and (3) how were they selected to be in the program? We return to these questions in Chapter 4 and present specific data collection procedures regarding these three types of evidence—eligibility criteria, measured characteristics, and the selection process. The issue of *representativeness* is important here. The evidence that needs to be marshalled can be found primarily in descriptions of eligibility criteria, enrollment procedures, and data summarizing relevant participant characteristics. These should be presented in the procedures or results section of the report or article.

It is important to look carefully at a program's eligibility criteria and enrollment procedures to determine who actually received services. In fact, we'll devote a major part of Chapter 4 to this topic. In the meantime, let's review the STETS eligibility criteria and recruitment procedures to provide an example.

Eligibility criteria for STETS were established for two purposes: to limit program participation to those who could potentially benefit from program services, and to encourage projects to recruit and enroll a broad range of clients in order to provide an adequate information base for examining the suitability of STETS for a diverse population. Thus, rather broad eligibility criteria were necessary to meet these two goals. Accordingly, each client was to meet the following criteria:

- Age between 18 and 24, inclusive.
- Mental retardation in the moderate, mild, or lower borderline range.
- No unsubsidized full-time employment of 6 or more months in the 2 years

preceding intake, and no unsubsidized employment of more than 10 hours per week at the time of intake into the program.
• No secondary disability that would make on-the-job training for competitive employment impractical.

Enrollment procedures are then described in the STETS report that summarize how projects were encouraged to recruit and enroll a broad range of clients in three ways. First, projects were expected to work with clients who had secondary disabilities, if those disabilities did not make the training and placement impossible or unreasonably difficult for the client or the project. Second, they were encouraged to enroll relatively more lower-functioning clients than they might have otherwise. And third, they were discouraged from enrolling only those whom they considered in advance to have the highest likelihood of success in the program (that is, they were discouraged from "creaming").

After the enrollment criteria were set out, a table of characteristics at enrollment was prepared. It was also noted that all persons that came into the program were referred by an agency. This was important because it might indicate that they all had some support for their vocational goals and someone to turn to as they entered and worked to stay in the labor market. It is this type of selection process that often presents a pitfall to consumers. They try to apply the results of a study to a group of persons who appear to resemble those served by an earlier program but because of undocumented differences in the selection process are actually different in unmeasured ways.

One should be able to find similar descriptions of the program's eligibility and recruitment procedures. The relevant question to ask is, if I were operating this program, what would I want to know about the applicants in order to design the appropriate treatment? Usually, those are the factors that should be covered in answering "who." The evidence that one then looks for or presents is actual descriptions of the persons served. These descriptions are most likely found in tabular format within the report. Our guideline here sets the standard for marshalling evidence for determining who was served:

Guideline 8. Evidence should be presented that documents who was served by the program and compares those persons with those originally projected.

Before leaving the issue of representativeness, let's discuss briefly the issue of "creaming," or working primarily with those participants that have the highest likelihood of success. The important thing to keep in mind is that the sample should reflect the target population, and that one needs to be thorough in looking for or describing who is actually served. For example, do participants in a program for the severely handicapped truly have severe handicaps; did the program take only those participants whose parents show an interest or those persons who can travel; did the program involve only a small segment of first-time offenders rather than recidivists; did the program enroll a cross section of handicapped students; or did the program involve only acute mental patients rather than both acute and chronic? These are frequently asked questions for which evidence should be marshalled to preclude a concern about creaming and/or nonrepresentativeness.

However, creaming is sometimes justified. For example, the Job Training Partnership Act demonstration includes a rather strict entry criterion because it is often limited to providing a very short (just a few days) intervention strategy. Thus, by design, it excludes persons who need more extensive services (for example, nonreaders). We feel that if creaming occurs, it should be clearly described in the setup and that generalizations should be limited to the sample group.

An additional aspect of "who" deals with attrition. It asks the question, do you have data on those people for whom you said you would have data, and were the data consistent over the entire period of the study? Attrition means that data sources and available information on people frequently change over time. Thus, if data are collected over time, data sources will frequently be lost; when that occurs, one really does not have complete data on all participants unless the attrition is random. If attrition is not random, one needs to worry about having a subset of participants who are not representative of the original sample group. In reference to the Long-Term Care project, for example, after 18 months, 39% of the original participants in the basic case management model had died. Although this doesn't suggest a differential attrition rate between groups, it does indicate that the remaining participants reflect a different group than that originally constituted. The finding also points out the importance of distinguishing between data attrition and program attrition:

- *Data attrition:* As time goes by (or for other reasons) you do not get data on all persons served. If the group with data differs from the others, you may have a problem.
- *Program attrition:* Not everyone enrolled in the program participates. Thus, the characteristics and impacts on all *enrollees* may differ from those for all *participants*. Again, systematic differences cause problems.

Our eighth guideline is still appropriate. Evidence should be available that discusses attrition and the effects it may have on the results. If the discussion is not in the report, be skeptical.

2. What

The evidence to marshall here pertains to describing the intervention provided to the participants. What were the services, what were their intensity and duration, and were they actually provided? Additionally, you need to describe the outcomes that resulted from the provision of services and the outcomes that would have occurred if the services were unavailable. Is your list of measured outcomes sufficient to enable you to form an opinion about the overall effectiveness of the intervention, or do you have partial information that will only support partial conclusions? This is an area where your tolerance for uncertainty will play a key role in telling you how many stones is enough—how big a foundation are you going to need to support the arguments and decisions that lead to your evaluation.

Marshalling the evidence regarding ''what'' includes two areas: (1) describing the intervention provided to the participants, including its intensity and duration; and (2) describing the outcomes that resulted from the provision of services and the outcomes that would have occurred if the services were unavailable.

a. Intervention Provided.   One should provide (or look for) indicators of such things as length of stay, costs, staffing patterns, placement rates, credits earned, hours in therapy, or hours of job support. Figure 3.1 summarizes the

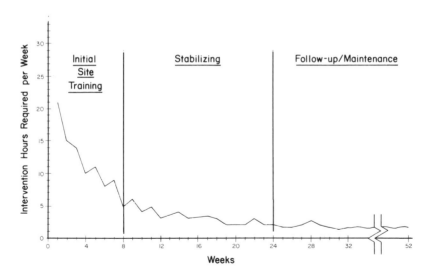

Figure 3.1. Hours of staff intervention required to maintain clients in competitive employment.

hours of staff intervention required to maintain mentally retarded clients in competitive employment (Hill & Wehman, 1985). You should find in the evaluation study or report similar evidence supporting the services or interventions actually delivered.

It is also necessary to provide evidence indicating that you measured what you said you would. Two examples will help set the stage for a later guideline. One pertains to a very popular concept, quality of life. The authors are familiar with a number of studies purporting to either change or improve a person's quality of life through the programmatic services delivered. Again, the consumer or provider must be skeptical, because there is little agreement in the literature regarding either the operationalization or measurement of quality of life (Schalock, Keith, Karan, & Hoffman, in press). Thus, measures of quality of life might pertain only to residential or working environments, rather than to more comprehensive measures that more adequately reflect a person's quality of life. More is said of this in Chapter 7.

A second example relates to data reflecting costs. It is very easy to fall into a trap in which only partial costs are presented. For example, in the longitudinal study of the court-ordered deinstitutionalization of Pennhurst residents (Ashbaugh & Allard, 1983), the costs to maintain persons in Pennhurst were significantly greater than the initially projected costs for community placement. On closer analysis, however, when the costs of providing day programs in the community were factored in, the actual costs were essentially equivalent. Similarly, if the program claims to affect total long-term costs, does it include *all* long-term care expenditures, or merely those covered by Medicare or Medicaid? Thus, one needs to ask, "costs to whom?" And if the program claims to have reduced overall expenditures, then the evidence marshalled should *justify that*, not a shifting of the costs from one governmental entity to another, or from the government to the person.

b. Describing the Outcomes on Both Sides of the Structured Comparisons. The outcomes measured depend upon the program's goals and objectives. They are also chosen on the basis of the setup that defines your data needs. Think back to the rationale of the setup that specifies the link among program participants, services received, and expected outcomes. As a producer or consumer, you will need to show (or see) that these outcome measures result logically from the program and that the measures capture important elements of comparison on both sides of the comparison (program participants and nonparticipants). A simple rule for you as the producer to follow is to include the information you would need if you were a consumer; that is, what would you want to know before you put your scarce resources into a program.

Generally speaking, we suggest that you stick with objective measures (which we discuss in detail in Chapter 7) that can be measured within the time period that you have, and that you ensure that the resources devoted to the process are appropriate to what you think you will get out of it. Thus, don't go overboard in measuring outcomes, for you still have to ensure that your program runs. The other thing to keep clearly in mind is the distinction between outcomes and impacts. Although we discuss this distinction in detail in Chapter 8, it is important at this point to remember that in marshalling the evidence, it is necessary to collect outcome data on both the treatment and control groups.

An example from the STETS demonstration reflects how outcomes should follow from the setup, and how one can describe the outcomes on both sides of the structural comparison. The purpose of STETS was to affect employment, use of alternate education and training services, and receipt of public transfers and to increase social independence. Most of the effects were assumed to stem from increased employment (as presented in the model in the setup), so the project focused most of its attention on that outcome. Exemplary outcome measures for both the experimental and control groups are presented in Table 3.4. The actual data in this sample table are not important at this time; what is important is to note that the measures adequately reflect their intended purpose, and that comparable data sets are provided for both the experimental and control groups. Hence, evidence was marshalled on both sides of the comparison.

Sometimes, however, these data must be obtained in a post hoc fashion. For example, in the Hill *et al.* (1985) study referred to in Chapter 1, there was no control group, so they marshalled the evidence for a ''post hoc group'' from published documents. In providing or evaluating the evidence marshalled, one needs to look closely at the assumptions made in using post hoc comparison data and how similar the post hoc group is to the program participants. There should also be comparable time periods for program participants and that reported for the post hoc group(s).

c. Conclusion. Thus, the ''what'' interrogatory focuses on marshalling evidence to describe the intervention provided and the outcomes that resulted from the provision of services. Thus our ninth guideline stresses the importance of these two functions:

Guideline 9. Evidence should be provided that describes the intervention provided, the outcomes that resulted from the provision of services, and the outcomes that would have occurred if the services were unavailable.

Table 3.4. **Exemplary Outcomes from Experimental and Control Groups (STETS Demonstration Project: Month 6)**[a]

Outcome measures	Experimental group mean	Control group mean
Employment		
Employed in regular job (%)[b]	11.8	10.7
Employed in any paid job (%)	67.8	45.2
Average weekly earnings in regular job	$11.81	$9.81
Average weekly earnings in any paid job	$52.39	$25.93
Training and schooling		
In any training (%)	61.7	40.6
In any schooling (%)	7.5	15.7
Income sources		
Receiving SSI or SSDI (%)	26.3	31.0
Average monthly income from SSI or SSDI	$66.41	$74.59
Receiving any cash transfers (%)	31.7	43.1
Average monthly income from cash transfers	$80.23	$99.98
Average weekly personal income[c]	$71.72	$50.94

[a]Adpated with permission from Kerachsky *et al.* (1985).
[b]Regular jobs are those that are neither training/work–study nor work-shop/activity center jobs.
[c]Personal income includes earnings, cash transfer benefits (AFDC, general assistance, SSI, and SSDI), and other regular sources of income.

3. Where/When

These two questions can often be addressed together. The issue is can you describe the context in which the intervention was fielded? Where did it take place, and when did it take place? Be sure to consider whether any special features of the location and time that might influence program performance are noted. The critical evidence you want to marshall in reference to "where" involves documenting the context of the evaluation, including the geographical location. For example, STETS was evaluated in five U.S. cities, whereas the Long-Term Care project was evaluated through 10 local projects distributed primarily in the Midwest and on the East Coast. The evidence for "where" does not need to be complex or detailed; it merely needs to clearly describe the context within which the program occurs. Is it a rural or urban program; is it offered as part of another program that might provide some of the support or direct services; is it offered as part of a demonstration or university-based program; or is it provided within the facility or outside in a more normal environment? This evidence reveals important factors that might affect outcomes, impacts, or replicability, and these are the issues readers will be (or should be) concerned about. This evidence becomes important in the next section on interpreting the findings within the program's context. Evidence regarding the time frame and period of operations is also important. For example, if it is an employment program, was it offered during a period of high unemployment—this was the case for STETS. If the evaluation was to look at long-term care costs, were truly *long-term* care costs evaluated or did the evaluation cover only 18 months, as did the Long-Term Care project. The importance of this interrogatory is reflected in Figure 3.1. During the initial site training, about 21 hours per week of staff intervention were required to maintain clients in competitive employment; after 24 weeks, only about 3 hours per week were required.

Impacts will be affected by the context within which the program was fielded. For example, employment is affected by the geographical area and the historical time frame. As we discuss in Chapter 5, there are ways to summarize contextual variables such as local labor market conditions, local infrastructure, transportation systems, and the like.

"Where" and "when" should not be overlooked in marshalling the evidence. Thus, we present our 10th guideline:

Guideline 10. Evidence should be provided that describes the context in which the intervention was fielded, when the services were provided, and for how long.

---

4. How/Why

The issue with respect to these two interrogatories is a *process* one: Can you describe how the intervention took place and why it had the effects you claim it did? These issues follow from the rationale developed in the setup and generally require that you review program operations to determine whether the elements identified in the rationale as being necessary are, in fact, present in the program being tested.

Marshalling evidence regarding the "how" interrogatory requires considerable data collection such as we describe in Section II on process analysis. Human service programs involve a number of intervention strategies or treatments based upon available resources, technology, and manpower. Examples include education, training, counseling, case management, assessment or evaluation, job placement, health care, residential services, transportation, and recreation. These interventions or services are provided by certain persons, in particular ways, over specific periods of time. The "what" interrogatory asks you to marshall evidence describing the services of interventions provided to the participants, including their intensity and duration; the "how" interrogatory ask you to marshall evidence regarding *what specifically was involved in the intervention or service*. For example, what was included in the education provided; what did training really include; which specific counseling techniques were employed; or what assessment or evaluation tools and techniques were actually used? This type of evidence is essential in being able to describe how the project or intervention accomplished its goals.

Marshalling the evidence regarding the "why" interrogatory requires answering the questions why you measured what you did, and did the measures really capture the important differences? It requires that you look at the issue or problem addressed and ask, why is it important to collect information on this issue, who needs this information, who is affected by this policy, and what does previous experience tell us? Remember our previous example of juvenile delinquents and the significant differences reported between delinquents and nondelinquents in milk consumed? In either conducting or reading such a study, one should look critically at the setup and specifically at the historical perspective and rationale presented that justify linking milk intake to delinquency. Similarly, in reference to the medication reduction study previously mentioned, in which the administrator of a mental hospital wanted to know the effects of reducing medication in psychiatric patients, why were the measures (behavioral change, restraint usage, and staff injuries) selected, and did they capture the important changes that occur with medication reduction?

And furthermore, you need to marshall evidence that demonstrates why the intervention had the effects you claim it did. Was there a relationship, for example, between amount of intervention and the outcome produced? Were higher (or better) outcomes found in those persons with greater amounts of education, training, or counseling? This is not a simple issue, as we see in the two sections on impact and benefit–cost analyses, for it involves data analysis and some degree of uncertainty. For the time being, you want to marshall evidence that describes how the intervention took place and explains why the intervention or program had the effects you claim it had. Thus, our 11th guideline:

Guideline 11. Evidence should be presented that describes how the intervention took place and why it had the effects you claim it did.

---

5. Conclusion

In summary, marshalling the evidence involves a number of guidelines governing program evaluation. Our four guidelines reflect the importance of evidence and imply rules that govern the evidence. But evidence must be interpreted in light of the program's purpose, context, and rationale. And it is to the interpretation of the findings that we now turn.

C. Interpreting the Findings

The basic question you want to answer in interpreting the findings is Are you convinced that the program produced the change, and if so, what's your degree of certainty? The goal of the producer here is to make this step easy for consumers. The importance of this section of the chapter is reflected in the medication reduction study referred to earlier. This was the study in which the administrator of a mental hospital wanted to know the effects of medication reduction and found they were not associated with significant behavioral changes, increased use of restraints, or increased injuries inflicted on staff by clients. But a skeptical audience came into play when the results were presented to the nursing and ward personnel, who were reasonably certain beforehand that medication reduction had deleterious effects. Thus, when it came time to interpret the findings, many staff were not convinced that the effects of the reduction were those found in the study. A producer encounters such an audience frequently; conversely, consumers frequently are that audience. Thus, as we now discuss guidelines regarding the interpretation of findings, it is important to keep both producer and consumer roles clearly in mind.

We discuss four guidelines regarding interpreting the findings. They include (1) addressing the correspondence between interrogatories; (2) determining whether the change can really be attributed to the intervention; (3) interpreting impacts within the program's context; and (4) assessing the limitations and weaknesses of the evaluation, including the degree of uncertainty in the results.

1. Correspondence between Interrogatories

If you did a good job linking the setup and marshalling the evidence, then this step is easy. You have information on the key outcomes under each situation (program and comparison), and you know about the intervening factors. Table 3.3 summarized the interrogatories concerning the setup and marshalling the evidence. As we discussed these in the previous section on marshalling the evidence, we essentially compared the two sets of interrogatories. For example, if you could not find the evidence suggested in each guideline, then as a consumer, you should be skeptical of the findings. Thus, if in comparing the "who" interrogatories you find that services were not provided to the persons specified, then you should doubt the study. If as a producer you find questionable correspondence between the interrogatories, it will be necessary to marshall more evidence before presenting the findings to that skeptical audience. That was exactly what was done in the medication reduction study, since the staff wanted to know other possible causes of the effects reported (such as medication change rather than reduction) before they were willing to accept the results. Whether you are a producer or consumer, the advantage of using Table 3.3 is that it makes interrogatory comparisons relatively easy as you attempt to interpret the findings.

2. Attributing the Effects to the Intervention

The question here is, can the change really be attributed to the intervention? The setup should have provided the structured comparisons, the program model, the expected outcomes, and the rationale linking services to outcomes. In marshalling the evidence, data on both sides of the structured comparison were collected. If all the above are in place, then your question is, do you think the difference is due to the intervention or something else? Some further questions you might ask will help answer this question:

- Are both sides of the comparison measured accurately, and are outcomes presented from the program participants and from those in a comparable situation?
- Did the measures used capture the important differences between the two scenarios?
- Are issues discussed such as internal validity, participant attrition, selection biases, differential participant history, and time-related changes that frequently and significantly reduce the program's ability to relate the effects to the intervention or services received?
- Were statistical tests used in the analysis of the data, and if so, were statistically significant differences reported? And further, were the significant differences presented in the context of sample selection, sample size, and amount of variation in the measured effects?

It is difficult in any evaluation to attribute the effects to only the intervention or services provided. Human behavior and the interventions used to change that behavior are too complex. In fact, this is probably the most dangerous area of evaluation, for all sorts of subtle reasons can account for why apparent differences in outcomes are *not* due to the program under evaluation. Although we cover this issue in detail in Chapters 8 and 9, you should keep the following points in mind:

1. The comparison made may not accurately correspond with the comparison defined in the setup. For example, in a pre–post comparison, the group may have changed from the "pre" state even without intervention (which is particularly true for employment for youth). Or, the comparison group might have been drawn from an area outside the catchment area for the program, thereby introducing different services and employment or health opportunities and constraints than the treatment group. Thus, observed differences are due to dissimilar local conditions, not the intervention.

2. Intervention may have been greater than you thought, and participants may get extra service not captured by the program definition. In reference to deinstitutionalization, for example, people in the community may get community mental health services. If these are not captured in the intervention definition, then wrong conclusions may result.

3. There may be bias in the data collection. This bias may result from systematic nonresponse for persons in one group or the other, so that comparisons are not between both groups but between one group and part of the other. Or, the bias can result from the use of different data sources, failure to control adequately for difference in the way data are recorded, or defining variables differently in different data sources.

Thus, we ask again that evaluators be logical and interpret their results within reasonable rules of evidence and scientific rigor. We underscore the importance of this request with the following caution:

Caution.   Make sure that the key outcomes are presented, that they are consistent with what is said in the setup, and that the differences in outcomes are attributable to the intervention.

---

3. Interpretation within the Program's Context

In interpretation the interrogatories of "where" and "when" come into play. What you want to focus on is whether the results are interpreted within the context of policies or geographical area that could potentially color the results. Some of the more important questions to ask in this regard include:

• Is this a demonstration program supported or conducted by a special provider who would not operate the replication program, or is it a regular ongoing program? Many demonstration programs offer special services (such as high staff–participant ratios, special context, or free general administrative support) that produce very different results than regular ongoing programs.

- Is it an urban or a rural program? Each venue poses its own problems and solutions, but the results need to be interpreted in light of the venue where implemented.
- Does the geographical area have a history of supporting the services provided? For example, different parts of the country historically have provided some services better than others. This is true in community mental health centers, programs for the developmentally disabled, employment programs, diversion programs, special education programs, and drug rehabilitation programs.
- What were local conditions affecting the outcomes during the project? For example, it is very important to include local employment rates in an employment program, since low employment or placement rates may well reflect a poor labor market rather than a poor program.
- Do the measures used capture the important differences? For example, in the Long-Term Care project, if you looked only at the Medicare costs you wouldn't capture all the costs, since many people spent their own money or were covered by Medicaid.

These five questions are not exhaustive, but it is hoped they indicate the need to interpret the results carefully and in light of the program's context. Again, logical reasoning will assist you in interpreting the findings regardless of whether you are assuming a consumer or producer role. But answering the questions will also help you to determine how far you can generalize the findings and what implications the results have for you. Can you apply your program, whose feasibility has now possibly been established, to similar participants in other areas? Thus, you should expect to be told (or tell others) how to go beyond this direct comparison and to what types of persons and environments replications are possible.

4. Interpretation within the Evaluation's Limitations

All evaluations and, indeed, all decisions involve uncertainty. Thus, any evaluation should address its limitations and potential weaknesses. All evaluations have limitations, and we ask that only state-of-the-art, comprehensive evaluations be done. Thus, we propose a guideline that reflects our caution and best advice:

Guideline 12. Results of any evaluation should be interpreted within the program's context and with a consideration of the evaluation's potential weaknesses.

---

5. *Conclusion*

In concluding this chapter on the guidelines regarding program evaluation, we feel that it is important to provide the reader with a helpful reference to the book's chapters and a self-survey regarding factors to look for or include in a program evaluation. The flowchart is presented in Figure 3.2 and the self-survey in Survey 3.1. If you are a consumer, Survey 3.1 lists what to look for; if a producer, these are factors to include in the evaluation.

III. Summary

In this chapter we addressed a number of guidelines regarding program evaluation that we feel apply to any type of program analysis including process, impact, and benefit–cost. The guidelines were presented within the analytical

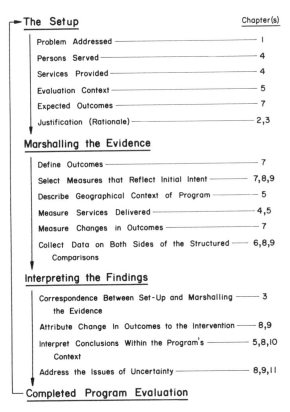

Figure 3.2. Program evaluation guidelines flowchart.

Survey 3.1. **Self-Survey**: **Factors to Look for or Include in a Program Evaluation**

Component	Factors to look for or include	Is the factor present?[a]	
		Yes	No
The setup	1. Problem addressed 2. Persons served 3. Services provided 4. Evaluation context 5. Expected outcomes 6. Justification (rationale)		
Marshalling the evidence	1. Outcomes defined 2. Measures selected that reflect initial intent 3. Geographical context of program described 4. Services delivered measured 5. Changes in outcomes measured 6. Data collected on both sides of the structured comparisons		
Interpreting the findings	1. Correspondence between setup and marshalling the evidence 2. Change in outcomes attributed to the intervention 3. Conclusions interpreted within the program's context 4. The issues of uncertainty addressed		

[a]Place a check mark beneath either Yes or No.

framework of the setup, marshalling the evidence, and interpreting the findings. The setup involves the evaluation's goal, context, and rationale that links the services provided to the expected outcomes. Marshalling the evidence, which flows from the setup and marshalling-the-evidence interrogatories, determines whether the program effects can be attributed to the intervention and evaluates the results in light of the program's context and limitations.

A number of guidelines were presented throughout the chapter including the following:

- The goal statement of your program evaluation should include (1) a well-defined purpose, (2) clearly defined structured comparisons, and (3) what you are trying to test with the data.

- Describe the context, particularly those special aspects that might limit generalizability, within which the evaluation is to be interpreted.
- Your setup should include the problem addressed, the persons served, the services provided, the evaluation context, the expected outcomes, and a rationale linking the intervention or services to the outcomes.
- Evidence should be presented that documents who was served by the program and compares these persons with those originally projected.
- Evidence should be provided that describes the intervention provided, the outcomes that resulted from the provision of services, and the outcomes that would have occurred if the services were unavailable.
- Evidence should be provided that describes the context in which the intervention was fielded, when the services were provided, and for how long.
- Evidence should be presented that describes how the intervention took place, and why it had the effects you claim it did.
- Results of any evaluation should be interpreted within the program's context and with a consideration of the evaluation's potential weaknesses.

We conclude the chapter with a caution regarding these guidelines; they are tightly interrelated, since the setup drives marshalling the data, which in turn results in the findings that must be interpreted. Thus, our caution:

Caution.   When you conduct a program evaluation, do not separate the setup from marshalling the evidence from interpreting the findings.

II

Process Analysis

We hope that after reading Section I on program analysis you more fully appreciate the importance of program evaluation and your potential role as an evaluation producer or consumer. You should also be aware of the guidelines governing program evaluation that tell you or your readers that the technical program evaluation criteria have been addressed. Throughout Section I we stressed a number of themes including the importance of using logical thought and the scientific method, the need for valid information to make good decisions, and the necessity of matching evaluation questions to available resources. We continue those themes in this Section on process analysis.

Process analysis describes the program and the general environment in which it operates, including the persons served, the services provided, and the costs involved. The importance of process analysis is that it provides potentially useful feedback to the administrator regarding the program and allows others to replicate the program should they feel it is feasible or attractive. Additionally, data resulting from one's process analysis provides the basis for conducting either impact or benefit–cost analyses.

In thinking about your role as a program administrator, there are probably three types of data that are most frequently requested or debated. These are data about who are the persons served, what services are provided, and what are the program's costs. These three questions are the heart of process analysis and provide the focus for this section's three chapters. Chapter 4, which focuses on the persons served and the services provided, addresses the participant's perspective in program evaluation. Participants are concerned primarily with the services received and their relevance to program progression and postprogram effects. Chapter 5 focuses on the program's perspective and addresses issues such as how the organization is set up and internal resources and external factors that influence the program. This perspective represents a considerable part of any administrator's day and is also a significant part of a process analysis. Chapter 6 addresses costs and includes subsections on our approach to cost analysis, perspectives on costs, and estimating costs to the program, government, and the participants.

Process analysis involves collecting and analyzing considerable data regarding such variables as participant characteristics, services delivered, or costs. Therefore, we discuss feasible data collection and analysis procedures within each chapter. We maintain our assumption that as a producer, you will most likely be involved in feasibility-stage process analyses. We want to familiarize you, as a producer or consumer, with an actual process analysis, and we use the Long-Term Care demonstration project to illustrate the various components of a comprehensive process analysis. The reader is cautioned that the complete Long-Term Care demonstration process analysis is nearly 700 pages long, and we are not advocating or suggesting such a criterion if you are a producer! We anticipate that the example will result in an appreciation of the value of a process analysis and the various components that should be contained therein.

Process analysis represents both a challenge and potential threat to program administrators. It challenges you to look at your program and collect data about who it serves, what it provides, and what it costs. That information can be used to your advantage for improving your program, introducing innovations, or being on top of what's going on so that your best defense is a good offense. It's potentially threatening because people are sometimes hesitant to present data that they will be held accountable for. This fear is compounded further by three realities we have frequently experienced. First, even though many programs do not do evaluation as we view it, they are sending out "program evaluations" through their public relations departments. Second, some evaluations may not look as good as the administrator had hoped, yet they still have to compete for money the next budget period. And third, many persons (including funders) are sometimes skeptical that you are truly using your program evaluation data for program improvement as opposed to political purposes. Thus, we anticipate that readers of this section will feel both emotions: excitement because of the challenge and apprehension due to the potential threat.

Process analysis does not deal with outcomes or impacts; rather, process analysis deals with program description. As a methodology, process analysis is very similar to the usual methodology employed in several published studies based on multiple case study designs (Greene & David, 1984; Shortell, Wickizer, & Wheeler, 1984). It reflects both the increasing use of rigorous case study designs and a trend toward the formalization of qualitative research. This trend has developed in response to several perceived research requirements, including the need to coordinate data collection at many sites, the need to facilitate cross-site data analysis, and the need to increase the reliability and validity of the findings. Additionally, we have incorporated into our proposed approach to process analysis the following characteristics of formalized qualitative research (Firestone & Herriott, 1983): an emphasis on exploration, the codification of questions and variables before beginning data collection, standardization of data collection procedures, and systematic reduction of verbal narrative to codes and categories.

Process analysis should be viewed as a helpful tool to any administrator for at least four reasons. First, it provides feedback regarding the program and its participants that can be used to improve the program. Second, it provides the necessary, detailed information about the program that allows others to replicate it if the program looks attractive. Third, it provides data for either clinical use or impact and benefit–cost analyses. And fourth, for the evaluation producer, it provides answers to many of the interrogatories concerning the setup and marshalling the evidence that are listed in Table 3.3. These include:

- Who: Did you serve whom you said you would, and from whom were the data collected?
- What: Did you deliver and measure what you said you would?
- Where: Did you document the context of the program, including when services were provided and for how long?
- How: Did you describe how the intervention took place?

4

Process Analysis from the
Participant's Perspective

I. Overview

Human service programs involve multiple types of services and participants. Thus, in this chapter we ask you to think about whom you serve and the services you provide. To accomplish this task, we need to focus on how participants were selected, measures that describe their relevant characteristics, and indicators of the program services they receive. Measurement is very important in this chapter, since the data can be used for a number of purposes. If you do impact analysis, for example, you will need data regarding participant characteristics and services received. You also might need to know about differential receipt of services, such as who came into the program and who graduated from it. Or you may want to conduct a process analysis wherein you describe how persons got into your program, their important characteristics, and what services they received.

Regardless of the need, we propose that the most helpful technique to use in generating the necessary data is to use a "table shell" such as that presented in Survey 4.1. This shell, or matrix, is a convenient and functional way to outline the necessary data, thus allowing you to think about how it can be collected and analyzed. The material presented in the top portion of Survey 4.1 lists two categories of participant characteristics evaluated in the National Long-Term Care demonstration project (hereafter referred to as LTC); the services listed in the bottom section of the survey are those common to many human service programs. The survey also identifies three alternative ways to aggregate the data. For example, for a multiphase program, one could collect and analyze the data across the different phases; if the project involved a structured comparison, data for the experimental and control groups would be collected and listed separately. The third option reflects cross-program comparisons. We present a number of table shells such as Survey 4.1 throughout the following chapters for expository and suggested use purposes.

Selecting which participant characteristics or program services to measure is

Survey 4.1. **Exemplary Table Shell for Planning Necessary Data Sets and How They Will Be Presented**

		Alternative ways to aggregate the data					
		Program phases		Structured comparisons		Multiple programs	
Data sets and how presented		Phase I	Phase II	Experimental	Control	Program 1	Program 2
Participant characteristics (percent of participants)							
Disability in activities of daily living (ADL)[a]							
Eating							
Transfer							
Toileting							
Dressing							
Bathing							
Impairments in instrumental activities of daily living (IADL)							
Meal preparation							
Housekeeping							

Transportation
Taking medicine
Money management
Telephone use

Services provided
    (hours/month)
Education
Training
Counseling
Case management
Assessment/evaluation
Job placement
Health care
Residential services
Transportation
Recreation

---

[a]Adapted from the National Long-Term Care demonstration project (Carcagno et al., 1986).

always a major issue to face and resolve. We feel that the resolution is easier if you keep the setup and rationale issues discussed in the previous chapter clearly in mind. However, there is still the issue of the time and money involved in data collection. Think about all the potential sources and techniques that you might use to collect data regarding your program and its participants. At a minimum, sources include the participants, their families, and other agency records (such as state departments, social service agencies, and the Social Security Administration). Collection procedures include interviews, extracts from service providers, or existing data systems. Thus, we feel that it will be helpful to present six criteria that are appropriate in assessing alternative data sources and procedures. These criteria, listed and defined in Table 4.1, should be self-explanatory. The six apply each time we talk about data from here on in the book.

The chapter is divided into three sections. The first focuses on the persons served and discusses how data can be generated to describe how participants were selected and their relevant characteristics. The second focuses on the services provided by your program and how you might document those services. The third discusses ways to analyze the data. As with the previous chapters, we ask that you use logical thought and interact with the text and exercises.

II. Persons Served

Characteristics regarding the persons served are central to replications and interpretations. Some of these characteristics can be measured, and some are implicit in the ways participants were selected. Hence, we begin this section with the question, how were participants selected?

A. Selection Procedures

The easiest way to describe your selection procedures is to draw a programmatic flowchart such as that presented in Figure 4.1. In developing your

Table 4.1. **Criteria for Assessing Alternative Data Sources and Procedures**

1. *Accessibility.* The ability or willingness of a data source to provide the necessary data.
2. *Completeness.* The extent to which data can be obtained for all sample members (sample completeness) and the extent to which a data source provides all the data necessary for the evaluation (data completeness).
3. *Accuracy.* How well the data reflect actual events and characteristics.
4. *Timeliness.* Whether the data pertain to the specific time periods of interest, and whether the data can be obtained without excessive delays after the occurrence of events under study.
5. *Flexibility.* The facility with which a data collection strategy can accommodate changes in research goals, budgets, or the general program environment.
6. *Cost.* The amount of money necessary to implement a particular strategy and the level of certainty with which cost projections can be made.

flowchart, pretend that you are a participant in your program. How did you learn about the program, and why did you decide to apply? Once you applied, how was your eligibility established, and what kind of screening or assessment was performed? For some applicants, part of the selection process is being offered the service and then turning it down. These persons, along with those who are not eligible or who voluntarily terminate, become nonparticipants and represent the status quo. This suggests that there are a number of decision points throughout one's program involvement; some are under the control of the program and some the applicant (see Figure 4.1). In reference to process analysis, the key thing to look for is whether people who come in are systematically different from those who do not. In addition, you can use the flowchart to determine whether there is a systematic "leakage" from the program and to realize that whom you serve depends upon where you look in the flowchart.

Selection procedures do affect a process analysis. In that regard, an analysis should address four questions: (1) what are the eligibility requirements, (2) what were the recruiting and outreach procedures, (3) what were the referral sources, and (4) how representative are the eventual participants? We use the LTC process

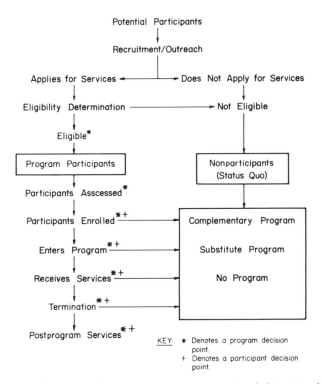

Figure 4.1. Flow chart reflecting selection procedures and typical program phases.

analysis (Kemper, 1986) to provide exemplary answers to each of these questions.

1. Eligibility Requirements

Human service programs generally have clearly stated eligibility criteria that an applicant must meet. For the LTC project, for example, the participant had to meet five criteria:

- Must be 65 or over.
- Must reside within the project catchment area and be living in the community (or if institutionalized, certified as likely to be discharged within 3 months).
- Must have two moderate ADL (activities of daily living) disabilities, or three severe IADL (instrumental activities of daily living) impairments, or two severe IADL impairments and one severe ADL disability (see Survey 4.1).
- Must need help with at least two categories of service affected by functional disabilities or impairments for 6 months or have a fragile informal support system that may no longer be able to provide needed care.
- Must be eligible for Medicare Part A (for the financial control model).

2. Recruiting and Outreach Procedures

Each program had its own approach to recruiting potential participants. Sometimes the procedure was quite formal, such as advertising in the media, and sometimes it involved only word of mouth. Additionally, programs dealing only with legal or statutory referrals may not need to do much recruiting or outreach. Commonly used procedures, such as those used in the LTC project, included formal outreach with existing providers, direct community outreach activities including involvement with the mass media, bulk mailings, and self-referrals. You may want to pause here and think about your program's recruiting and outreach procedures.

It is important to describe one's eligibility criteria and recruiting procedures, but it is also necessary to describe how potential participants are screened for program involvement. In reference to the LTC project, for example, applicants for services came to the attention of the screening unit as self-referrals, referrals from agencies, or referrals from family, friends, or neighbors. Screeners were instructed to conduct the interviews directly with potential participants whenever possible, but they could also accept reports from formal referral sources, families, friends, or other proxies, as necessary. The interesting fact was that almost 60% of the screening interviews involved some proxy participation. This is an important piece of information, since it suggests something about the participants' physical condition and motivation for participating. It might also be used later in impact or benefit–cost analyses to help explain the obtained results.

3. Referral Sources

Referral sources say something about both the selection procedures and the environment within which the program operates. In reference to our example, referrals were received from over 20 types of sources, including medical equipment suppliers, apartment and hotel managers, clergy, relatives, hospitals, and local service agencies. These multiple referral sources were aggregated into three primary categories: informal sources, health-related sources, and social service agencies. Once referral data were aggregated to one of these three categories, they became part of the process analysis and provided valuable data for it. For example, 28% of the referrals came from informal sources such as family, friends, or self; 42% were from health-related sources; 20% from social service agencies; and 10% from other sources.

4. Representativeness

A key question to answer regarding your selection procedures is whether the people who came into the program are systematically different from those being served by other programs. Examples of nonrepresentativeness are plentiful, such as the LTC project, which took only people who were referred or volunteered; the STETS project, whose participants all had to be within the referral system; drug programs wherein one must be referred by clinics or physicians; community-based programs for the mentally retarded in which one must be SSI eligible; and the Job Corps, which recruited only from specific areas.

One way to evaluate your participants' representativeness is to compare them on a number of factors from other data sources. This was the approach used in the LTC project's process analysis. By using previously published data, the evaluators were able to compare the participants' characteristics with those reported in other published sources. As one might expect, some similarities and differences were found that provide important descriptive and interpretive data. An example can be seen in Table 5.5.

5. Conclusion

In summary, it is important to describe your selection procedures so that you can describe clearly how participants get into your program and how they compare with nonparticipants. The relevant question to ask in this regard is, are the participants systematically different? This question is important throughout our discussion of process, impact, and benefit–cost analyses. Thus, we propose our 13th guideline:

Guideline 13. Understand your selection procedures clearly so that you can determine whether people who come into the program are systematically different from those who do not.

B. Participant Characteristics

One of the cardinal rules in feasibility-level evaluations is that to replicate a program, you must clearly describe who is served by that program. This raises the question, which participant characteristics should I describe? One possibility is to use the criteria listed in Table 4.2; another is to focus on what information *you* would want to know about the participants in a program you wished to replicate? In regard to that question, we propose three criteria for choosing which participant characteristics to describe:

1. Choose characteristics you can measure.
2. Choose those that you think will strongly affect the outcomes.
3. Choose those that have variation that will permit demonstrating an impact.

1. Measurable Characteristics

An important part of a process analysis is to develop a broad range of participant descriptors. These descriptors can be used for a number of purposes that range from simply describing the major characteristics of program participants, to indicating whether appropriate persons are being served, to assessing and comparing reported program outcomes. Participant characteristics can be divided into several categories such as those listed in Table 4.2. The three categories listed include basic demographic data, descriptors of the participant's abilities, and previous work, living, and service experiences.

The menu of descriptors listed in Table 4.2 is not a comprehensive list, and no single program will use all these data sets. In addition, data from each of the three categories are generally used in other ways, and not just for a process analysis. For example, basic participant demographics are the global measures used to describe the current participants. Ability data are generally used to describe variations in participants' abilities across programs, to determine whether the persons selected for services match the stated program goals, or to

Table 4.2. **Measures of Participant Characteristics**

Basic demographic data
    Age
    Gender
    Ethnicity
    Legal status
Abilities
    Activities of daily living
    Developmental level/quotient
    Adaptive behavioral level
    Educational level
    Health-related conditions
Experiences
    Residential history
    Family involvement
    Work history/income
    Education history
    Public assistance history
    History of services received
    Institutional involvement: mental
       health, corrections, mental
       retardation

evaluate the level of program outcome taking into account the characteristics of the persons served. Data related to experiences are important because they reflect the range of program participants and can be used as baseline measures to track individual changes through programmatic involvement. Hence, the participant's previous work, living, and service histories can be used to identify participant longitudinal characteristics that might influence programmatic outcomes and provide pre–post programmatic comparisons. Data from either the abilities or experiences categories can also be used to determine the clinical needs of participants.

You will undoubtedly use more than one participant characteristic in describing the persons served by your program. Regardless of the number of descriptors used, each should fulfill the basic criterion of measurability.

An important part of a program's process analysis is to summarize the baseline assessment level of the participants. This baseline measure permits one to determine the comparability of comparison groups before programmatic involvement. An example from the LTC project is presented in Table 4.3. Although it is a detailed listing, the tabled data allow us to make a number of points about measured participant characteristics. First, note the structured comparisons involving the basic case management and financial control models. These comparisons appear quite similar at baseline. Second, note the comprehensiveness of

Table 4.3. **Exemplary Measurable Participant Characteristics**[a]

Baseline assessment characteristics	Basic case management model	Financial control model
Basics		
Mean age (years)	79.2	80.1
Ethnic background (%)		
Black	22.3	23.0
Hispanic	2.0	3.7
White or other	75.6	73.3
Education (%)		
None	5.4	5.4
Elementary/high school	83.5	82.8
Some college	11.2	11.7
Female (%)	71.9	70.6
Married (%)	31.9	32.9
Abilities		
Mean number of ADL disabilities[b]	2.7	2.8
With ADL disablity (%)	16.6	15.8
Incontinence (%)	52.5	53.6
Mean number of IADL impairments[b]	5.5	5.5
With IADL impairments (%)	0.5	0.2
Mobility impairment (%)	80.1	82.8
Impairment in mental functioning (%)		
None or mild impairment	30.3	28.7
Moderate mental impairment	38.3	38.9
Severe mental impairment	31.4	32.4
Self-perceived health (%)		
Excellent	16.4	18.1
Fair	29.5	32.8
Poor	54.2	49.1
Experiences		
Type of residence at baseline (%)		
LTC facility	3.7	1.4
Hospital	8.2	15.1
Supportive housing	1.3	2.1
Private residence	86.8	81.4
Usual living arrangement (%)		
Alone	35.1	39.1
With spouse or spouse & child	31.1	31.7
With child (no spouse)	22.7	19.0
With other	11.0	10.2
Mean monthly income	$567.	$572.
Owns home (%)	44.7	38.9
Experienced stressful life event in past year (%)	86.0	87.4
Loneliness (%)		
Often lonely	27.0	25.7
Sometimes lonely	33.6	36.4
Never lonely	39.2	38.0

Table 4.3. (*Continued*)

Baseline assessment characteristics	Basic case management model	Financial control model
Experiences (*continued*)		
Social contacts	9.4	10.2
No contact in past week	6.3	7.2
One contact in past week	27.8	27.1
2–6 contacts in past week	56.6	55.4
Daily or more frequent contact	3.3	4.0
Mean number of unmet needs (maximum 8)[c]		
Satisfaction with life (%)	20.8	15.4
Completely satisfied	39.7	37.2
Pretty satisfied	39.5	47.4
Not very satisfied		
Attitude toward nursing home placement (%)	7.3	6.3
Wait-listed or applied for nursing home	63.4	67.3
Unwilling to go into nursing home		

[a]Adapted from the Long-Term Care demonstration process analysis (Kemper, 1986).
[b]See Survey 4.1.
[c]Need categories included bathing, dressing, toileting, transferring, medication preparation, housekeeping, transportation, and medical treatments.

the participant characteristics assessed. Although there may be more characteristics than appropriate for your program, a reasonable standard is to use two or more from each of the three categories presented in Table 4.2. And third, these characteristics are measurable. Although not presented in Table 4.3, assessment instruments are also listed and described in the LTC project's process analysis.

2. Characteristics That Affect Outcomes

Our second criterion for choosing which participant characteristics to look at is to choose those you think will strongly affect the outcome(s). We refer to these characteristics as *conditioning factors*. At this point go back to Chapter 2 and your setup. Look at your program model, the expected outcomes, and the rationale that explains why you think that *these* services provided to *these* participants will produce *these* outcomes. For example, if you have an employment program, what are the participant characteristics that might logically affect the outcomes? Possible characteristics include intelligence level, job history, education, family support, and mobility. Similarly, the outcomes from mental health or correction programs may be affected by participant characteristics such as previous family and institutional involvement, work and residential history, and educational level. These conditioning factors are the characteristics that need to be measured and reported as your participant characteristics.

In reference to the LTC example, the setup suggested in part that a managed approach to providing community-based long-term care could maintain or improve the well-being of clients. Thus, the participant-referenced outcomes of most interest to the project included longevity, reduced functional deterioration, improved social and psychological well-being, lower unmet needs, and increased service satisfaction. Those participant characteristics listed in Table 4.3 had been shown (or can logically be seen) to affect the outcomes.

A quick exercise will be helpful here. Go back to Chapter 2 and review your setup, including your program model, expected outcomes, and rationale. Develop a table shell such as presented in Survey 4.2, and fill in the baseline assessment characteristics for your program and, if possible, the comparison program. If it is impossible to use an experimental–control group design, such as that in Table 4.3 (the structured comparisons of the basic case management and financial control models), there are other means of identifying a comparison group. Such a comparison group should resemble the group under study and conform with the specified comparison situation (see Table 3.1, "the setup"). For example, if the comparison situation is a specific alternative program, the analyst would look for a group that was being served by that program. Other comparison

Survey 4.2. Table Shell regarding Participant Characteristics and Structured Comparisons

Baseline assessment characteristics[a]	Your program	Comparison group
Basics (fill in)		
1.		
2.		
3.		
4.		
5.		
Abilities (fill in)		
1.		
2.		
3.		
4.		
5.		
Experiences (fill in)		
1.		
2.		
3.		
4.		
5.		

[a]See Tables 4.2 and 4.3.

groups might include persons who were offered program services but who did not actually enroll, or persons in similar circumstances in other geographical areas where the special service is unavailable.

There is a problem, however, with these approaches to generating a comparison group: the treatment and the comparison group may differ in ways that confound one's ability to isolate the effects of the intervention. For example, differences in individual characteristics, economic and social opportunities, parental support, ability, or motivation may create differences between the treatment and comparison groups that have nothing to do with the program impacts we discuss in Section III. Thus a caution:

Caution. If a comparison group is used other than one generated through random assignment to experimental–control conditions, be sure to include rigorous statistical controls for intergroup differences that might influence the outcomes of interest.

---

3. Variation in Conditioning Factors

The third criterion for choosing which participant characteristics to describe is to choose those that have variation in the conditioning factors that will permit demonstrating an impact. For example, if all persons have the same (or maximum) score on the conditioning factors measured, there is no way to demonstrate an impact. Note in Table 4.3 that in reference to either the percentages or mean scores presented, there is considerable variation both between and within the two groups. This variation is an essential requirement for demonstrating impacts as well as for meeting the requirements of many of the statistical procedures referenced in Sections III and IV.

4. Conclusion

In summary, relevant participant characteristics provide the basis for all three types of program analyses, including process, impact, and benefit–cost. In reference to a process analysis, participant characteristics are essential for descriptive and replication purposes. A guideline regarding which ones to look at and use is given below.

Guideline 14. Choose only those participant characteristics that you can measure, that you think will strongly affect the outcomes, and that have variation which allows you to demonstrate an impact.

C. Data Collection

Now that you know which participant characteristics you will use, the next question is, how do I set up the data system to get the data? You have lots of options, but the method that we recommend is to include the desired characteristics on the intake or application form. An example is presented in Figure 4.2.

III. Services Provided

Human service programs involve a number of intervention strategies or treatments based upon available resources, technology, and manpower. Before considering the specific services you provide, let's think about how you could use service data in a process analysis. You might be interested, for example, in how many participants drop out or how many receive assessment, counseling, or job support services. These data sets are easy to generate and do not require sophisticated data systems. In fact, you can probably get these data by talking to your staff. If, however, you are interested in the intensity of your services or what happens over time, a more sophisticated data system will be necessary. Such a system would need to have the capability to provide a record of specific services delivered to specific participants over a predetermined time period. You can collect these data, but it does require more time and expense. The advantage of the more complex approach is that it allows you to be more complete in describing the services provided to your program participants. Additionally, impact and benefit–cost analyses frequently require data regarding the intensity and duration of services.

Thus, at the outset of documenting services provided, you must decide what questions to address, which data sets to collect, what resources are required to collect the data, and whether the benefits of a more complex analysis outweigh the resources involved in collecting the data. As a consumer, you are in an enviable position, since someone else has already collected the data for your readership. As a producer, you might want to follow our 15th guideline:

> *Guideline 15.* Data collected regarding services provided should be chosen on the basis of the questions addressed in the analysis and whether the benefits resulting from a more complex analysis outweight the costs of collecting the additional data.

A. Selecting the Services

Return to Survey 4.1 for a moment. In the bottom section of the survey we listed a number of services typically offered in human service programs. These include education, training, counseling, case management, assessment/ evaluation, job placement, health care, residential services, transportation, and recreation. You may wish to pause here and develop a table shell listing the services you provide and how you are measuring those services. Typical measures include hours, days, or units of service.

The list of services in your table shell should be as detailed as necessary—but don't go overboard. Keep the services clearly defined, easy to measure, and related to your outcomes. A maximum of five or six is probably a good rule, unless you need clinical-level information or are involved in a large-scale process analysis such as the LTC project. Their process analysis involved monitoring receipt of the following services:

Housemaker/personal care	Chore services
Skilled nursing	Mental health
Home health aide	Adaptive and assistive equipment
Home-delivered meals	Respite care
Therapies	Day maintenance
Companion services	Other (noncare)
Transportation	Adult foster care
Housekeeping	Housing assistance
Day health	Emergency assistance
Medical supplies	

We discuss data collection forms in the next subsection. First, however, we need to present a guideline regarding selecting which services to document:

Guideline 16. Be sure that the services provided and documented are easily defined and measured and logically affect your outcomes.

B. Documenting the Services

You should now know the services you provide and for which you will provide documentation. We hope you have followed our 15th and 16th guidelines. The next question becomes, how do I collect the data? Again, you have a number of options, and we discuss three.

You could use a client service record such as that provided in Figure 4.3. The record is built around six services provided by a transitional employment training program. Staff members involved in the project merely fill in the date, their name, time taken per service, and then circle the service provided. Data from the record could be used to summarize the amount of services received by each participant, along with the service provided and date of service.

	2. Applicant's Home Phone Number	3. Social Security Card Available? ☐ YES ☐ NO	Annualized Income Total $

P E R S O N A L   I N F O R M A T I O N

4. Applicant's Name (Last, First, Middle Initial)		5. Social Security Number

6. Home Address	7. City	8. State	9. Zip Code	10. County	11. Message Phone

12. Person to Notify in Emergency	12a. Relationship	12b. Address	12c. Telephone Number

13. Applicant's Age	13a. Birthdate month day year	14. Ethnic Group: ☐ White ☐ American Indian ☐ Asian or Pacific Islander ☐ Black ☐ Hispanic	15. Citizen of U.S.? ☐ YES ☐ NO If NO, Enter I 9 4 #

15a. Permanent Resident Alien? ☐ YES ☐ NO	15b. If a non Citizen, Do You Have a Work Permit? ☐ YES ☐ NO	16. Are You a High School or College Student? ☐ YES ☐ NO	17. Last Grade Completed?

18. Family Members Employed by JTPA? ☐ YES ☐ NO	19. Number in Household	19a. How Many Household Members Do You Support?	19b. Are You Claimed as a Dependent on Your Parents' Income Tax Form? ☐ YES ☐ NO

Family Income

List all members dependent on family income including yourself. Use additional sheet if necessary

I N C O M E   I N F O R M A T I O N

19c. Family Member	19d. Age	19e. Relationship	19f. Source of Income and Employer's Name	19g. Estimated Income for Last 6 Months	19h. Estimated Income for Last 12 Months
				$	$

* If Family Income Totals Zero, Attach an Explanation of Your Means of Support — *** TOTAL FAMILY INCOME** $

20. Farm Income? ☐ YES ☐ NO	NOTE: To qualify as a farm, gross annual sales of farm products must be $1,000 or more.

S O C I A L   S E R V I C E S

21. Are You Receiving Any of the Following? ☐ SSI ☐ ADC ☐ Food Stamps ☐ Other _____	22. Displaced Homemaker? ☐ YES ☐ NO

23. Employed? ☐ YES ☐ NO 24b. Registered Date month day year	24. If Not Employed, Give Date of Unemployment	24a. Job Service Office Location	A displaced homemaker is someone who has not been in the labor force and who has lost their main source of income—i.e. widow, recently separated, divorced or loss of public assistance.

25. Are You Receiving Unemployment Benefits? ☐ NO ☐ YES If YES, Give Date Benefits Began month day year	26. Have You Received a Lay-Off Notice Within The Last Six Months? ☐ NO ☐ YES If YES, Give Date month day year

Figure 4.2. Exemplary intake/application form (adapted from Job Training Partnership Act application).

EMPLOYMENT RECORD

27 List Last Four Jobs, The Most Recent First:

Name and Address of Employer | Job Duties (list machines, materials and equipment used and the number of people supervised)

Dates of Employment (Month, Day, Year)
From: To: Average # of Hours Worked Per Week
Salary/Wage Per Week $ Your Job Title
Reason for Leaving

Name and Address of Employer | Job Duties (list machines, materials and equipment used and the number of people supervised)

Dates of Employment (Month, Day, Year)
From: To: Average # of Hours Worked Per Week
Salary/Wage Per Week $ Your Job Title
Reason for Leaving

Name and Address of Employer | Job Duties (list machines, materials and equipment used and the number of people supervised)

Dates of Employment (Month, Day, Year)
From: To: Average # of Hours Worked Per Week
Salary/Wage Per Week $ Your Job Title
Reason for Leaving

Name and Address of Employer | Job Duties (list machines, materials and equipment used and the number of people supervised)

Dates of Employment (Month, Day, Year)
From: To: Average # of Hours Worked Per Week
Salary/Wage Per Week $ Your Job Title
Reason for Leaving

VETERANS AND OTHER INFORMATION

28 Veteran's Status ☐ YES ☐ NO | 28a. Dates of Military Service From: To: | 28b. ☐ Vietnam Era ☐ Special Disabled | 28c. Discharge Other Than Dishonorable ☐ YES ☐ NO

29 Limited English? ☐ YES ☐ NO | 30. Refugee? ☐ YES ☐ NO | 31. Do You Have Any Handicap That Has Affected Past Employment or May Hinder You In The Future? ☐ YES ☐ NO If YES, Explain:

32. If You Turned 18 On or After January 1, 1980, Have You Registered For The Draft? Federal Law Requires Registration For The Draft.
☐ YES ☐ NO
Selective Service Number: _____

33. Are You Currently a Regular Outpatient of a Mental Hospital or Mental Health Clinic?
☐ NO ☐ YES If YES, Give the Name and Location

34. Are You a Legal Offender? Any Adult or Juvenile Who Is or Has Been Subject to Any Stage of the Judicial Process.
☐ YES ☐ NO | 34a. Are You Now, or Have You Been Within the Last Six Months, Confined in a Hospital, Prison, Sheltered Workshop, or Similar Institution?
☐ NO ☐ YES If YES, Give the Name and Location

Figure 4.2. (*Continued*)

Figure 4.3. Client service record. Service codes are A—evaluation, B—preplacement training, C—job development, D—on-the-job services, E—off-the-job services, and F—follow-up.

A slightly more complex format is presented in Figure 4.4. This format allows one to collect considerable data regarding services provided. However, its use does involve considerable resources.

Longer process analysis evaluations, such as the LTC evaluation, may need a fairly elaborate participant tracking system. The client tracking form from the LTC is presented in Figure 4.5. In addition to that shown in Figure 4.5, a second form was submitted each time there was a change in the participant's status. This form is presented in Figure 4.6.

Documenting the services you provide will require resources; therefore, we again stress the importance of our 16th guideline, which states, ''be sure that the services provided and documented are easily defined and measured and logically affect your outcomes.'' The sample forms we included have been used to provide the data discussed in the chapter. But the forms are not the entire story; additionally, one must consider how they are used. For example, the status-change form shown in Figure 4.6 was used only when an LTC project participant reported a status change. This involved less paperwork, but might have been less consistent with the goals of some programs than if the participant's status was reported daily. Daily status reports would allow less chance for overlooking a status change, but more resources would also be used. Thus, the data collection involved in documenting services is always a trade-off between needs and resources. Our best advice is to be very clear about the questions asked in the analysis and whether the service data collected will answer those questions. If not, then probably the data are not important and do not warrant the resources involved in their collection.

IV. Data Analysis

Now that you have collected data on the persons served and the services provided or, as a consumer, have that data in front of you, the question becomes, what do I do with it? Answering that question is the focus of this last part of the chapter. We base our discussion on a simple premise: all analysts start out by looking at correlations or relationships in the data. We further suggest that it is probably sufficient initially to use a cross-tabs, data-based systems approach to your data analysis. The proposed cross-tabs approach represents the first ''cut'' of the data, but it will undoubtedly provide you with valuable data, suggestions, and insights. Additionally, if you think carefully about the results, you can learn a great deal. Even if you are doing (or reading about) more complex analyses, we suggest that you start with a simple two-dimensional cross-tabs approach.

Our discussion of data analysis is based on Figure 4.7, which summarizes data from a recent analysis completed by Robert L. Schalock. The analysis involved developmentally disabled participants who were enrolled in a transi-

Participant S.S. #: _____   Month: _____, 19 ____

Participant need level: high, moderate, low   Component (Check): CL ____

Staff I.D. #: _____   ES ____

                                                                                       CI ____

Service unit type	Specific activity/objective (limit 10 characters)	Service intensity (units provided per day)							Total units for month
		1	2	3	4	5	--------	----31	
Training 1									
2									
3									
4									
5									
Assistance 1									
2									
3									
4									
5									
Support 1									
2									
3									
4									
5									

Figure 4.4. Individual program plan-referenced units of service recording sheet. One unit of service intensity equals 15 minutes of direct services. The program components are CL—community living, ES—employment services, and CI—community integration.

tional employment program. The participants' need status was eigher high, moderate, or low, as described in Schalock and Keith (1986). The portion of the analysis discussed here relates to attrition rates and differential receipt of services.

The left side of Figure 4.7 shows that the same number of high-, moderate-, and low-need participants completed the assessment phase. Thus, attrition did not differ among the need levels during the project's first phase, which verifies that the participants received the assessment service. This type of analysis can be done with a very simple data system that allows one to relate (cross-tabulate) the participants' need level to completing the assessment phase.

However, one might also be interested in a more detailed question such as, did they make it through each stage of the program? This question regarding differential attrition involves a more detailed data system, but the same cross-tabs approach to the analysis. The center section of Figure 4.7 summarizes the number of participants who completed each of the three programmatic phases, including assessment and work readiness (I), transitional jobs (II), and postplacement support services (III). As can be seen, there was a differential attrition during the second and third phases for moderate- and high-need participants. When you see this in your data, don't stop with the cross-tabs; rather, go back to the data about participant characteristics and see potentially why the differential attrition occurred. For example, we found that high-need participants, who were most likely not to make it through Phases II and III, tended to be older males who were more severely mentally handicapped and who had entered the training program later than the other need-status groups. The importance of this additional analysis is that it permits you to develop some hypotheses about why these persons did not complete all phases. It may be due to the above characteristics, or it may be the model is incorrect for the high-need person. But don't allow yourself to be fooled; the cross-tabs approach shows only relationships, not causality. Thus our 17th guideline:

Guideline 17.   Cross-tabs analysis should not be treated as demonstrating causality; it only shows relationships.

The right side of Figure 4.7 summarizes the actual amount (measured as a unit of service, or 15 minutes of staff time) of training (''teach me''), assistance (''help me''), and support (''do it for me'') received by the different participant need levels during Phase II. This type of analysis is both more complex and informative; but it can still be done using a cross-tabs approach. The question that the data on the right side of Figure 4.7 addresses is, who gets what services? As can be seen, high-need participants receive more services during Phase II than

[__ __] [_____] [__]   CLIENT TRACKING FORM
        MPR APPLICANT/CLIENT ID#
                                                                                3/29/82

CLIENT/APPLICANT INFORMATION   WORKER IDENTIFICATION

NAME:_____   SCREENER: [__]__]__]__]
PERMANENT ADDRESS:_____   ASSESSOR: [__]__]__]__]
            _____   1ST CASE MANAGER: [__]__]__]__]
TELEPHONE:_____   2ND CASE MANAGER: [__]__]__]__]
BIRTHDATE: [__ __]__ __]__ __] (Mo/Day/Yr)   EFFECTIVE DATE: [__]__]__]__]__]__]
PROXY NAME:_____   1ST REASSESSOR: [__]__]__]__]
PROXY TELEPHONE:_____   2ND REASSESSOR: [__]__]__]__]
REFERRAL SOURCE:_____   EFFECTIVE DATE: [__]__]__]__]__]__]

ACTIONS   DATE   OUTCOMES *   REFERRED TO *   DATE REFERRED
                 [Month/Day/Year]   [Circle One]   [Mo/Day/Yr]

                                 SEND COPY OF TF TO MPR   o Reason Inappropriate
                                 FOR BOLD-FACED OUTCOMES   at Screen (Circle One)

I. SCREENING   1 Too Service Dependent
• A. REFERRED TO   [IAM_T IAO_T IAY_T]   2 Insufficient Disability
       UNIT. . . . .[__ __]__ __]__ __]   3 Insufficient Unmet Need
                                                                  4 Age Under 65
• B. SCREENING   [IB1M_T IB1O_T IB1Y_T] [IB2_T]   5 Outside Catchment Area
       INTERVIEW   6 Not Medicare Eligible
                                 1. INAPPROPRIATE   7 Other
                                    [Circle Reason o]
                 [__ __]__ __]__ __]. 2. REFUSED   _____[__] [__]__]__]__]__]__]
                                                     REFERRED TO
                                 3. UNABLE TO
                                    COMPLETE (OTHER)
                 [__ __]__ __]__ __]. 4. APPROPRIATE

   C. SUPERVISORY
       REVIEW. . . .[__ __]__ __]__ __]
• D. RANDOMIZA-   [ID1M_T ID1O_T ID1Y_T] [ID2_T]   _____
       TION   [__ __]__ __]__ __]. 1. CONTROL
       DECISION   _____[__] [__]__]__]__]__]__]
       RECEIVED   REFERRED TO
                   [__ __]__ __]__ __]. 2. CLIENT

   E. SCREEN SENT
       TO MPR. . . .[__ __]__ __]__ __]

   F. CONTACT
       ASSESSMENT   COMPLETE SECTION V IF CLIENT
       UNIT. . . . .[__ __]__ __]__ __]   DROPS OUT AFTER RANDOMIZATION

   G. SCREEN & TF
       TRANSFERRED TO
       ASSESSMENT   [__ __]__ __]__ __]

                                        (1)

Figure 4.5. Client tracking form for long-term care demonstration.

MPR CLIENT ID # [__ __] [__ __ __ __ __] [__] CLIENT NAME _____

ACTIONS	DATE	OUTCOMES •	REFERRED TO •	DATE REFERRED
	[Month/Day/Year]	(Circle One)		(Mo/Day/Yr)

II. BASELINE ASSESSMENT

SEND COPIES OF TF TO MPR FOR BOLD-FACED OUTCOMES

• A. ASSIGNED TO
WORKER. . . .(__ __)__ __)__ __)

[IIAM_T IIAD_T IIAY_T]

• B. OBTAIN
INFORMED
CONSENT   (__ __)__ __)__ __)

[IIB1M_T IIB1D_T IIB1Y_T   IIB2_T]   _____   [IIB3_T]

1. REFUSED +
2. UNABLE TO   ( )__)   ( )__)__)__)__)__)
COMPLETE (OTHER) +   REFERRED TO

(__ __)__ __)__ __). 3. COMPLETE

• C. BASELINE   [IIC1M_T IIC1D_T IIC1Y_T   IIC2_T]   _____   [IIC3_T]
ASSESSMENT

1. INAPPROPRIATE +
(__ __)__ __)__ __). 2. REFUSED +   ( )__)   ( )__)__)__)__)__)
REFERRED TO

3. UNABLE TO
COMPLETE (OTHER) +

(__ __)__ __)__ __). 4. APPROPRIATE

D. ASSESSMENT
SUMMARY FORM
COMPLETED . .(__ __)__ __)__ __)   + Complete Section V.A.

E. SUPERVISORY
APPROVAL. . .(__ __)__ __)__ __)

III. CARE PLANNING

A. ASSIGNED FOR
CARE PLAN
PREPARATION. (__ __)__ __)__ __)

• B. CARE PLAN
COMPLETED   [IIIBM_T IIIBD_T IIIBY_T]
(INCLUDING
SUPERVISORY
APPROVAL). . (__ __)__ __)__ __)

• C. CARE PLAN [IIIC1M_T IIIC1D_T IIIC1Y_T   IIIC2_T]   [IIIC3_T]
APPROVED
BY CLIENT/
FAMILY   (__ __)__ __)__ __). 1. REFUSED +   ( )__)   ( )__)__)__)__)__)
REFERRED TO

(__ __)__ __)__ __). 2. ACCEPTED (ACTIVE)

[IIIDM_T IIIDD_T IIIDY_T]
• D. FIRST SERVICE
INITIATED. . (__ __)__ __)__ __)

E. COPY OF TF
SENT TO MPR. (__ __)__ __)__ __)

IV. ARRANGING/MONITORING/REASSESSMENT   SERVICES ADDED OR DELETED

FIRST REASSESSMENT   REASSESSMENTS
SCHEDULED FOR: _____   COMPLETED: | DATE | COMMENTS |

NEXT SCHEDULED: _____   _____

| CLIENT TRACKING FORM | (2) | MPRI 738 |

Figure 4.5. (*Continued*)

MPR CLIENT ID # [___ ___] [___ ___ ___ ___ ___] [___] CLIENT NAME ____________________

V. STATUS CHANGES AFTER INDIVIDUAL BECOMES CLIENT

EFFECTIVE DATE OF STATUS CHANGE: Mo./Day/Yr.	VAM_X VAD_X VAY_X [__ __][__ __][__ __]	[__ __][__ __][__ __]	[__ __][__ __][__ __]	[__ __][__ __][__ __]
CHANGE STATUS FROM:TO (Circle One)	FROM: TO VB_X VC_X	FROM: TO	FROM: TO	FROM: TO
1. ACTIVE:	1 1	1 1	1 1	1 1
2. INACTIVE:	2 2	2 2	2 2	2 2
3. TERMINATED:	3 3	3 3	3 3	3 3
REASON INACTIVATED OR TERMINATED (Circle One)	VD_X			
1. INSTITUTIONALIZED:	1	1	1	1
2. REFUSED	2	2	2	2
3. MOVED OUT OF AREA:	3	3	3	3
4. CONDITION STABILIZED:	4	4	4	4
5. UNABLE TO LOCATE:	5	5	5	5
6. DIED:	6	6	6	6
(GIVE DATE OF DEATH)	VEM_X VED_X VEY_X [__ __][__ __][__ __]	[__ __][__ __][__ __]	[__ __][__ __][__ __]	[__ __][__ __][__ __]
7. INSUFFICIENT DISABILITY/ UNMET NEED:	7	7	7	7
8. TOO SERVICE DEPENDENT:	8	8	8	8
9. NOT MEDICARE ELIGIBLE:	9	9	9	9
10. REFUSED COST SHARING:	10	10	10	10
15. OTHER (SPECIFY):	15 ________	15 ________	15 ________	15 ________
REFERRED TO:	VF_X [__][__]	[__][__]	[__][__]	[__][__]
(SPECIFY)	________	________	________	________
DATE SENT TO MPR (Mo./Day/Yr.)	[__ __][__ __][__ __]	[__ __][__ __][__ __]	[__ __][__ __][__ __]	[__ __][__ __][__ __]

VI. APPEALS

DATE	STATUS	COMMENTS
______		____________________
______		____________________
______		____________________
______		____________________

Figure 4.6. Client tracking update form for long-term care demonstration.

either moderate- or low-need participants; yet, we just saw that they have a higher attrition rate. This finding is significant in and of itself, and would be used in the process analysis to answer the "what" interrogatory from Table 3.3, did you deliver what you said you would? Additionally, this result would affect the impact analysis, since you cannot show impacts until you show that the groups got the same intervention. Thus, both outcomes and impacts are affected by the participant's need-status level.

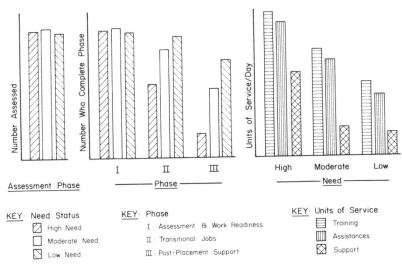

Figure 4.7. Examples of different ways to analyze participant-referenced data.

Because results such as the above are fairly common, asking questions about causation will quickly take you beyond the scope of the book. There are, for example, multilevel cross-tabs approaches, regression analyses, and analyses of variance that permit more complex data analysis. Additionally, we recognize that human services represent a multidimensional problem, not easily reduced to a two-dimensional cross-tabs approach. But we do feel that for most administrators reading this book, the suggested cross-tabs approach probably gives you enough information to do a respectable process analysis.

We would like to conclude our discussion of data analysis with a caution regarding sample size. Many human service programs are small, and therefore some of the cells in the cross-tabs may contain a very small number of participants. We caution you against acting prematurely on these data. In reference to Figure 4.7, for example, the center section of the graph suggests that high-need participants are differentially dropping out. But, if one looks at the right side of Figure 4.7, it is apparent that they are getting more service once they get into Phase II. Our caution will hopefully prevent you from the pitfall of acting prematurely on the results of your cross-tabs if you have a small number in a paticular cell:

Caution. Any time you have a small cell count (under 10) in your cross-tabs, treat it as a potential fluke; do not act prematurely without additional data or analyses.

V. Summary

In this chapter we focused on process analysis from the participant's perspective. We have discussed the persons served, including how they were selected and their characteristics. Additionally, we discussed the programmatic services offered by a program and how those services might be documented. Finally, we suggested an approach to analyzing data regarding persons served and services provided that involves a reasonably simple two-dimensional cross-tabs approach.

The material and techniques discussed in the chapter were presented within the context of a process analysis, which, in total, describes the program and the general environment in which it operates, including the persons served, the services provided, and the costs involved. Process analysis involves collecting and analyzing considerable data. Thus the chapter also discussed the use of various approaches to data collection and suggested six criteria for assessing alternative data sources and procedures including accessibility, completeness, accuracy, timeliness, flexibility, and cost.

As with the other chapters, we also presented a number of guidelines:

- Understand your selection procedures clearly so that you can determine whether people who come into the program are systematically different from those who do not.
- Choose only those participant characteristics that you can measure, that you think will strongly affect the outcomes, and that have variation which allows you to demonstrate an impact.
- Data collected regarding services provided should be chosen on the basis of the questions addressed in the analysis and whether the benefits resulting from a more complex analysis outweigh the costs of collecting the additional data.
- Be sure that the services provided and documented are easily defined and measured and logically affect your outcomes.
- Cross-tabs analysis should not be treated as demonstrating causality; it only shows relationships.

Collecting and sharing data about the persons served and the services received represents a challenge to program administrators. It challenges you to look at and evaluate your participants and the services they receive. Chapter 5 helps explain how those services are provided.

VI. Additional Readings

Denzin, N. K. (1970). *Sociological methods: A sourcebook*. New York: McGraw-Hill.
Lansing, J. B., & Morgan, J. N. (1971). *Economic survey methods*. Ann Arbor, MI: Institute for Social Research, University of Michigan Press.

Murphy, J. T. (1980). *Getting the facts: A field work guide for evaluators and policy analysts.* Santa Monica, CA: Goodyear Publishing.

Shortell, S. M., Wickizer, T. M., & Wheeler, R. G. (1984). *Hospital–physician joint ventures: Results and lessons from a national demonstration in primary care.* Ann Arbor, MI: Health Administrator Press.

Stainback, S., & Stainback, W. (1984). Methodological considerations in qualitiative research. *Journal of the Association for Persons with Severe Handicaps, 9*(4), 296–303.

Sorensen, J. E., & Elpers, J. R. (1978). Developing information systems for human service organizations. In C. C. Attkisson *et al.* (Eds.), *Evaluation of human service programs* (pp. 127–140). New York: Academic Press.

Process Analysis from the Program's Perspective

I. Overview

Our focus in Chapter 4 was on the persons served and the services provided by your program. In Chapter 5, the focus is on how those services are provided, and how services are affected by the program's organizational structure and a number of internal and external factors. Again, you will need to either generate or evaluate considerable data in your role as a producer or consumer. But the data in this chapter tend to be more qualitative and descriptive than they were in Chapter 4. Their primary uses in a process analysis are descriptive and explanatory: descriptive in that you need to describe the program and the environment within which it operates, and explanatory in that you can use the data to evaluate whether the intervention or services were implemented as designed and are generalizable to other situations.

We suggest that you begin thinking about process analysis from the program's perspective by referring back to Chapter 3 and the rationale that links the (programmatic) services provided to the expected outcomes. That rationale is shown diagrammatically in Figure 5.1. Now you need to focus on the *specific programmatic services* and ask, what will impact or affect that process? Just as in the last chapter we looked for participant characteristics that affect outcomes, here we analyze internal and external programmatic factors that potentially affect the outcomes. For example, labor market conditions, such as unemployment rate, may affect the outcomes of an employment program, whereas the crime rate within a program's catchment area may affect a correctional program's recidivism rate. The point we wish to stress is that human service programs occur in an environment that affects them in many ways. A process analysis should include a description of the most salient aspects of that environment and an explanation as to how environmental factors might affect the program's implementation and services delivered. This notion of a program's current environment is diagrammed in Figure 5.2.

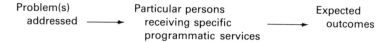

Figure 5.1. Rationale linking programmatic services to expected outcomes.

Some of the material in Figure 5.2 you saw previously in Figure 4.1, the flowchart reflecting selection procedures and typical program phases. In this chapter, we focus on information outlined in the center of Figure 5.2: a description of the *program;* a number of *internal factors* such as philosophy and goals, phases of programmatic development, and resources; and three *external factors*—formal linkages, relevant community descriptors, and family characteristics. As in Chapter 4, we use a number of examples from the National Long-Term Care (LTC) demonstration (Carcagno *et al.,* 1986) to demonstrate how the particular data can be used in a process analysis.

A basic question that runs throughout the chapter is, how much information is really required to describe and explain how the services were provided within the context of the program's current environment? Indeed, entire books are devoted to this topic. Our general rule is to remember the mother-in-law test and the KISLF acronym—keep it simple, logical, and focused. Our specific guideline suggests:

Guideline 18.   Describe your program and its environment in enough detail to allow its replication.

II. Organization Description

Human service programs represent multifaceted organizations that vary on a continuum from reasonably simple to unbelievably complex. To help you understand that complexity and to help us describe "The Program," we developed Table 5.1. The table contains five descriptive areas ("descriptors") that should facilitate achieving the intent of the 18th guideline regarding replication. Most human service programs are legitimated and monitored through either a governmental or corporate structure, and their governance structure includes a board of directors. Funding sources are multiple. Funding dimensions explained typically focus on the certainty of funding and/or control over financial incentives. Major service components usually include evaluation or assessment, training, education or rehabilitation, case management, and follow-along. Examples and data sources for each descriptive area are also presented in Table 5.1.

Before considering an example from the LTC project, let's pause so you can think about your type of organization, governance structure, funding source(s)

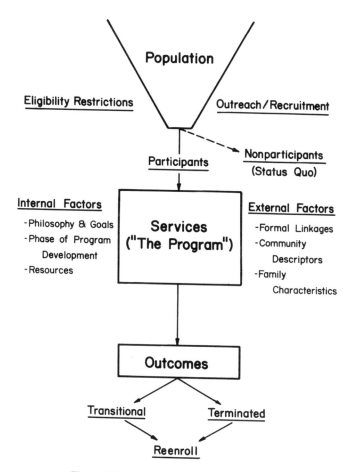

Figure 5.2. A program's current environment.

and dimensions, and major service components. We have provided a table shell in Survey 5.1 that we hope will assist you in your thinking. Fill in the survey based on what you know about your program and the material presented in Table 5.1.

Our example of organizational design and functioning from the LTC project is a bit longer than previous examples for at least three reasons. First, research evidence suggests that organizational design affects measures of organizational performance. Second, you will need to provide some detail regarding your organizational design and functioning for replication purposes. And third, we feel the ensuing material represents a good model that may be useful to you.

As you might remember, the LTC demonstration tested whether a managed

Table 5.1. Organization Type, Governance Structure, Funding Considerations, and Major Service Components

Descriptor	Examples	Data sources
Type of organization (legal designation)	Government: includes municipalities, state or federally operated facilities, city–county programs/facilities, etc. For-profit corporation: ordinary business corporations created and operated to make a profit that is distributed to the owners (shareholders). Includes private rehabilitation programs. Not for-profit corporation: includes charitable, educational, (re)habilitation, recreational, social, and similar organizations.	Incorporation documents Annual report
Governance structure	Board of directors Executive council Committee structure	Organizational chart Administrative policies and procedures
Funding considerations Funding source	Taxes Grants Foundations Client use fees Bonds	Budget Audit Annual report
Funding dimensions	Certainty of funding Control over financial incentives	Longitudinal analysis of preliminary versus final budget Trended analysis of staff salaries Participant payroll (if rehabilitation program)
Major service components	Evaluation or assessment Training, education, or rehabilitation Case management Follow-along	Agency brochures Personnel policies Policies and procedures manual Descriptions

Survey 5.1. **Table Shell regarding Organizational Type, Governance Structure, Funding Aspects, and Service Components**

Descriptor	Fill in for your program	Data source you used
Type of organization		
Governance structure		
Funding source		
Funding dimensions		
Major service com- ponents		

approach to providing community-based long-term care could help control costs while maintaining or improving the well-being of its clients and their informal care givers. The primary product of the channeling projects was the provision of case management services to frail, elderly clients. With certain federally imposed restrictions on the ways in which the projects could organize themselves to provide these services, considerable variation in organizational design was permitted. Figure 5.3 provides an overview of the organizational design and functioning of a hypothetical channeling project. Two cautionary notes are in order before we discuss Figure 5.3. First, the figure relates to the organization of the host agency, which was only part of the overall project. Indeed, there were a number of federal, state, and evaluator agencies also involved. And second, the figure depicts roles or functions of the channeling project; it is not an organizational chart. In practice, some of these roles or functions were combined in a single position.

The case managers occupied the fundamental and indispensable positions in the channeling project. They were responsible for developing care plans, arranging for services, monitoring service delivery, and adjusting care plans as necessary. When the channeling projects were fully operational, the number of case managers ranged from 4 to 10, and the client-to-case-manager ratio from 36 to 54. Over time, 8 of the 10 projects chose to develop positions for case aids, who functioned primarily in support of the case managers by helping with paperwork, ordering services, and monitoring clients.

Evaluation needs required the establishment of a screening unit administratively separate from assessment and case management in order to limit the contact between channeling staff and control group members, and thereby avoid control group contamination. This separation typically was accomplished by subcontracting the screening function to a different agency or housing it in a different physical location within the host agency under separate supervisory personnel. The role of the screeners was to administer a standardized survey instrument to potential clients and use its results to determine whether applicants met the acceptance criteria.

The process for purchasing services for channeling clients differed between the two channeling models. Financial control projects were able to reimburse services from a funds pool established by authorizations of state governments and the federal government. Basic case management channeling projects had a much more limited ability to purchase services using gap-filling funds. Under each model, it was necessary that case manager purchase requests and provider invoices be processed and reconciled when necessary. A subunit within the channeling project generally performed this function. Financial control projects were also required to maintain an automated financial control system, standardized across projects, to provide data for a variety of purposes. To accomplish this task, all financial control projects created the position of data manager and/or data clerk.

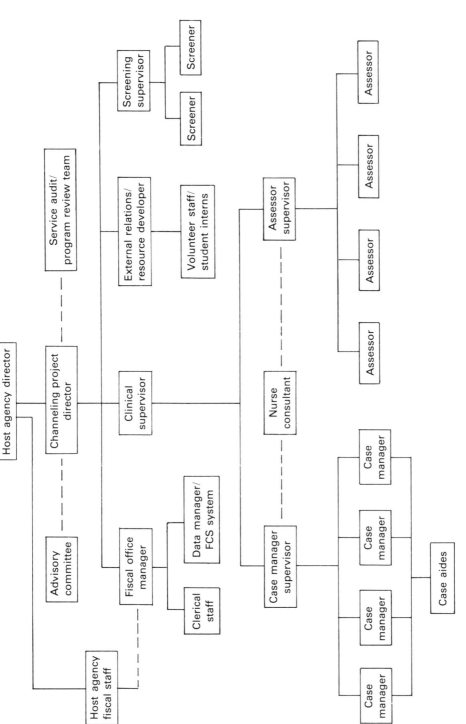

Figure 5.3. Hypothetical channeling project functional design.

In order to arrange for direct services for clients and to generate referrals of potential clients, it was necessary for each channeling project to develop and maintain ongoing relationships with community providers of acute and long-term care and to establish the project's visibility within the community as a whole. This responsibility typically was shared among several project staff members.

Two additional elements in the channeling project's organizational design included project advisory committees and a service audit and program review (SAPR). The advisory committees typically provided input to site operational procedures and were also used extensively to assist in case finding and working with referral sources. The SAPR team's purpose was to evaluate the programmatic performance of the project and provide suggestions to project management for improvement.

We hope the above description provides a general picture of how the channeling projects were organized to carry out their various functions. Your description may be different, based upon how you completed Survey 5.1. But regardless of the description you provide, we do hope that Figure 5.2 and the related material present an adequate working model to describe your organization and major service components.

III. Internal Factors

We have found that a number of internal factors can influence the process and outcomes of a human service program. Three factors, listed in Figure 5.2, are philosophy and goals, phase of program development, and resources. Each is discussed in this section, along with examples from the LTC project.

A. Philosophy and Goals

Human service programs today operate from philosophical concepts such as normalization, mainstreaming, crime control, equal opportunity, public health, income support, education, wellness, competitiveness and full employment. The importance of clearly stating the philosophy of a program cannot be overemphasized: one's philosophy provides the basis for the program's goals, service delivery pattern, and expected outcomes. For example, assume that a rehabilitation program's philosophy can be translated into those goals, services, and expected outcomes shown in Figure 5.4. Since one's philosophy drives the program, it should be described in any process analysis.

B. Phase of Program Development

There are a number of ways to conceptualize different stages or phases of program development. In Chapter 1, for example, we suggested that most read-

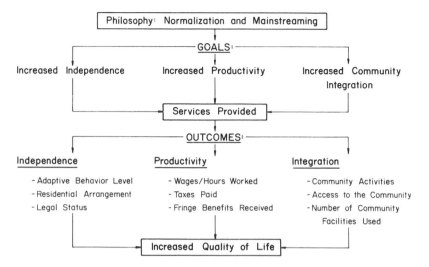

Figure 5.4. Relationships among philosophy, goals, services, and outcomes.

ers are concerned about the feasibility stage of program development, in which evaluation focuses on demonstrating that the program is promising or feasible. In this chapter, we suggest that programs can have different *phases* of development, whether they are in the feasibility, demonstration, or ongoing stage. In process analysis, it is important to describe the program's developmental phases and then to specify from which phase or phases the data were collected. An example from the LTC project demonstrates the importance of this point.

The channeling demonstration involved 3 federal agencies, 10 states, 10 local sites, a technical assistance contractor, and an evaluation contractor. The organizational structure of the demonstration was necessarily complicated, for it involved many government agencies at the federal, state, and local levels as well as the host agency (previously discussed in reference to Figure 5.3). There were four chronological phases to the project. The *demonstration planning phase* lasted approximately 18 months, during which sites were selected, detailed administrative and operational procedures designed, staff hired and trained, relationships with local providers and referral sources established, Medicare and Medicaid waivers obtained, financial arrangements for reimbursement of providers completed, data collection instruments designed, and research procedures affecting the local projects defined. The *buildup phase,* also lasting approximately 18 months, was characterized by intense outreach and screening efforts to build project caseloads. There were two distinct periods within this phase. The randomization period, which generally started when projects began accepting clients, was the period during which applicants were randomly assigned to the treatment or control group. Following this, the residual buildup period started, in which the projects added clients to their caseloads to meet the preestablished caseload targets. During this period, the projects also adjusted their staffing patterns to accommodate the requirements of the steady state phase. During the *steady state phase,* which lasted for 9 months, the projects were required to maintain a steady caseload size in spite of continual turnover in the caseload. This phase most nearly resembled the operation of an ongoing program. The last phase, *demonstration closeout,* lasted for 10 months, during which projects stopped accepting new clients and implemented their plans for closeout. Evaluation data did not cover this phase.

We included this example to make several points. First, it describes very well the complexity of program development and evaluation, since the project was almost constantly in a state of transition. Second, the project really did not generate evaluation data except for about 18 months during the latter part of the buildup and entire steady state phases. And third, a process analysis should contain a description of all such phases, since the phase structure and duration will obviously affect how and when programmatic services are provided.

C. Resources

The third important internal factor affecting a program is its resources, defined broadly to mean money, staff, experience, and time. A discussion of these resources is important to include in a process analysis, since they both describe a program and influence its outcomes. For example, it is important to know whether the program involves a big facility that has accumulated considerable equipment or is a mom-and-pop organization. Similarly, is the program affiliated with a large university that provides technical support and expertise? Data sources for obtaining this information include budgets, personnel records, interagency agreements, program narratives, and time studies. We use two examples that reflect the importance of including resource data in a process analysis. The first example reflects how staff functions might be explained in a process analysis and the importance of these data to the analysis. The second example demonstrates a wide discrepancy between what staff are hired to do and what they actually do.

We referred previously to the critical role case managers played in the LTC project. They came from a variety of professional backgrounds, including social work, nursing, psychology, behavioral disabilities, and sociology. They had either masters' or bachelors' degrees plus experience in either social work or counseling. All projects had one case management supervisor or two, typically with masters' degrees in human services, most often social work. Their two major functions were (1) initial care planning and service arrangement and (2) ongoing case management. Specific tasks within each function are presented in Table 5.2. In the project's process analysis, there is a detailed description of each task.

Table 5.2. **Case Management Functions and Tasks from the LTC Project**

Initial care planning	Ongoing case management

1. The care-planning process
  a. Review assessment data[a]
  b. Review client information
  c. Prepare care plan
  d. Select service providers
  e. Integrate informal service providers
  f. Involve physicians
2. Cost control elements of care planning[b]
3. Care plan review
4. Initiating and arranging services
5. Purchasing services
	1. Contacts with clients and providers
	2. Monitoring
	3. Client reassessment
	4. Termination of clients

[a]See Table 4.3 for representative assessment areas.
[b]Guidelines stipulated that for the entire caseload of the financial control projects, average expenditures could not exceed 60% of the combined Intermediate Care Facility/Skilled Nursing Facility Medicaid nursing home reimbursement rate. Projects could allocate these dollars across caseloads with some flexibility.

The process analysis should contain not only a description of staff functions and tasks, but evaluation by the staff regarding the procedures used. For example, in the LTC project, the specific forms and associated tasks were consistently criticized for excessive length, client burden, volume of paperwork, and lack of flexibility. This was reported to reduce direct contact with clients below desired levels. Many case managers and administrative staff also indicated that an automated system of care planning would be useful in helping to eliminate much of the manual accounting and cost calculation work undertaken by case managers. In the financial control system particularly, this cumbersome process was not helpful in providing information back to case managers which they could use for practice.

Considerable elapsed time between referral and screening and initial service delivery (over 1 month on average) was also identified as a problem by a number of referral sources. Delay in initiating services experienced by channeling seemed excessive. For example, some clients who were screened and found eligible did not receive an assessment for 2 to 3 weeks. Delays like this meant that clients judged to be at risk received no attention from a case manager or assessor to determine whether or not their immediate needs were being met by other means. This delay could have caused problems, particularly in those cases in which a crisis that could result in nursing home placement was the factor which caused the referral to channeling. Thus, it appears that channeling suffered from a lack of flexibility to meet occasional immediate needs while the care-planning process was going on. These are important observations and data to include in the process analysis, especially in light of the 18th guideline regarding replication.

The second example regarding resources demonstrates what program administrators frequently find: a wide discrepancy between what staff are hired to do and what they actually do. A 30-day time study was done (by Robert L. Schalock) in a rehabilitation program whose administrator was concerned about the lack of formal training being provided to persons with developmental disabilities. The program was funded on the basis of five participants to one instructional staff, ostensibly hired to provide training. The administrator assumed that this 5:1 ratio should result in considerable training being done.

The actual study involved having three types of instructional staff record in 15-minute time blocks the specific staff function performed. The staff functions and results for community living, employment services, and community integration instructors are presented in Table 5.3. Needless to say, the administrator was surprised at the results, which explain in part why he wasn't seeing more training being done. This result suggests the need for our 19th guideline:

Guideline 19. In reference to internal factors, never assume that what staff are hired to do is what they actually do.

In summary, internal factors do impact a program and thereby influence its outcomes. The three that we have identified and discussed here—philosophy and goals, phases of program development and resources—are not the only ones. Yet our experiences have indicated that they are worth your consideration and should be included in a process analysis.

IV. External Factors

If you refer back to Figure 5.2, you will see that we have listed three external factors—formal linkages, community descriptors, and family characteristics—that influence a program. The reader is cautioned that this is an area within which it is easy to get lost or mired. If you think about *all* the external factors that can affect your program, both your frustration and anxiety will increase. There are probably an endless number of economic, social, and political factors that can affect your program, and if you try to list them all, two things may well happen: you will fail the mother-in-law test, and your eyes will gloss over. Thus, we would like to begin this section by considering the criterion regarding external factors that is our 20th guideline:

Table 5.3. **Time Spent in Performing Different Staff Functions**

Staff function[a]	Time spent by type of instructor (%)		
	Community living	Employment services	Community integration
Appointments	0.3	0.1	0.3
Assessment	0.3	0.2	3.2
Assistance	29.4	16.1	21.5
Data-based training	15.0	16.5	3.9
Data entry	0.3	0.8	0.0
Fiscal	1.0	0.4	0.0
General maintenance	7.4	1.2	1.5
In-service training	2.5	1.5	3.4
Individual program plan (development & meetings)	2.8	3.2	12.5
Meetings (non-IPP)	3.3	4.4	6.2
Personal	4.0	8.7	8.0
Personnel	0.3	0.3	1.1
Public relations	0.8	0.1	10.0
Quality control	2.1	29.2	0.0
Recording	6.0	5.5	5.8
Scheduling	0.3	0.0	1.9
Search and rescue	0.2	0.0	0.0
Supervision	12.6	7.2	5.2
Support	8.2	1.9	12.0
Transportation	2.8	2.5	3.5
Typing	0.4	0.1	0.0

[a]Based on 150 staff.

Guideline 20. Consider only those external factors the environmental effects of which exceed some threshold and thereby can reasonably be expected to affect the program and its outcomes.

The guideline suggests not all factors that theoretically might affect a program actually do so. For example, assume you are an administrator of a diversion program funded by the Department of Corrections. What are some external factors that might realistically affect your program and its outcome? The communitie's crime rate, unemployment rate, population density, number of educational and recreational facilities, available job-training programs, and degree of family involvement might meet the criteria in Guideline 20. The same external factors might affect similar programs such as chemical dependency, employment training, and mental health. For an education program, external factors such as tax base, school board policy, family involvement and support, nutritional programs, and interfacility sharing agreements might be the significant factors to consider. Thus, once again you should use logical thought to identify only those external factors the environmental effects of which exceed some threshold and thereby can reasonably be expected to affect the program and its outcomes.

In sensitizing you to potential external factors, we discuss three that we feel impact most programs—formal linkages, relevant community descriptors, and family characteristics. We again draw on the LTC project and other sources for appropriate examples.

A. Formal Linkages

One of the realities most human service program administrators face is that, despite some persons' expectations, their program singularly cannot provide all necessary services. Frequently the wisest course of action is to join with other agencies or groups to increase the array of services provided. Describing how you accomplish formal linkages is an important part of a process analysis. The LTC project, for example, devoted four chapters in its final report to the development and maintenance of agreements with service providers.

A few years ago, one of us published an article entitled, "Comprehensive Community Services: A Plea for Interagency Collaboration" (Schalock, 1985). The article discussed the current trend toward establishing interagency and intersector linkages and then outlined the major phases involved in developing and maintaining linkages. These phases are listed in Table 5.4 along with suggested topical areas to include in a process analysis to summarize your successes and failures.

There is no established format to this section of a process analysis. The LTC project, for example, devotes one chapter to a discussion of developing and maintaining agreements with the formal service providers who were reimbursed for their services. The chapter includes a description of the factors influencing the selection of reimbursed formal service providers, the establishment of unit prices, and the bidding procedures used with the various entities involved. A second chapter discussed the monitoring procedures and standards used to improve service delivery. A third chapter discusses the use of individually contracted service providers, who were persons not part of the client's immediate family and who agreed to provide a service or a set of services to a client for a specified salary that exceeded out-of-pocket expenses. And a fourth chapter discusses linkages and the use of informal care givers and volunteers as service providers.

Formal (or informal) linkages are probably a part of your program. Think for a moment about all the agencies and service entities with whom you interact.

Table 5.4. **Stages in the Development of Linkages and Suggested Topical Areas to Include in a Process Analysis**[a]

Linkage phase	Topical areas
1. Determine feasibility	How did you learn about the services the entity could provide? How were initial contacts made? What were the incentives to each party?
2. Choose linkage model and membership	Which of the following common collaborative models was selected: (a) shared services; (b) contractual services; (c) centralized planning and administration? Why was one model chosen over another? Who composed the membership; who were "linked"?
3. Develop an operational plan	What types of agreements/contracts were entered into? What administrative arrangements were involved?
4. Operate and monitor the linkage	Who assumed responsibility, and for what? What type of adminstration policies and procedures were implemented? What sorts of performance indicators were used to monitor the "welfare" of the linkage?
5. Evaluate and change if necessary	What were the preestablished goals and outcomes for the linkages? If change was necessary, how was it accomplished? If termination was required, how was it accomplished?

[a]Adapted from Carcagno *et al.* (1986) and Schalock (1985).

They undoubtedly represent an important external factor to your program. How you describe the linkages is up to you: you may wish to use the phases presented in Table 5.4, or you may wish to use an interrogatory approach. But in your role as either a producer or consumer of evaluation, formal linkages should be described in a process analysis.

B. Relevant Community Descriptors

No human service program occurs in isolation; rather, it is a part of a larger environment that has specific economic, social, and political characteristics.

This section of the chapter asks you to think about those characteristics, their descriptors, and which descriptors will meet our criterion: to exceed some threshold and thereby affect the program. Across program types, relevant community descriptors include:

- Attitudes regarding human services (such as reflected in historical funding patterns).
- Auxiliary support services (such as education, mental health, rehabilitation, diversion programs, medical, and nutrition).
- Crime rates.
- Economic indicators (such as average income and employment rate).
- Public health indicators (mortality, life span, or disease rates).
- Tax structure and policies.
- Transportation availability.

The importance of examining the environments in which a program operates relates to the potential influence local conditions can have on the measurement and analysis of the program's impacts. In addition, describing the environment provides evidence about how the program's site(s) compares to a larger area such as a county, state, or the nation. In reference to the LTC project, for example, a sample of 10 sites selected through a competitive procurement process cannot be expected to yield a statistically representative sample of communities; it is nevertheless useful to examine the relevant community descriptors of the sites and make a judgment as to whether, taken as a group, they are broadly similar to the country as a whole.

We suggest that the rationale and procedures used in the LTC project and process analysis represent a valid model to follow as you think about relevant community descriptors to include (or look for) in a process analysis. We realize that your resources are limited; thus, use the following material as a model, not as a standard.

The LTC project's process analysis included two relevant community descriptors: (1) demographic and economic characteristics and (2) long-term care and related services in the demonstration areas. The demographic and economic characteristics of the aged population can be expected to affect the channeling intervention in important ways. For example, the size of the pool of potential referrals could have affected the speed of building caseloads; or the economic status of the population could affect both the proportion eligible for means-tested programs (such as Medicaid) that provide some community long-term care services and the proportion of the population able to purchase community services with their own resources. Given the rationale, data are then presented to compare the sites with the nation on relevant demographic and economic descriptors. We present some of these data in Table 5.5 to serve as a suggested model for your analysis.

Table 5.5. **Exemplary Relevant Community Descriptors from the National Long-Term Care Demonstration**

Descriptor	Basic case management mean	Financial control mean	United States
Demographic characteristic			
Total 65 and over (thousands)	74	117	25,543
Proportion of total population (%)	10.6	15.8	11.3
Female (%)	59.9	61.5	59.6
Living arrangement (%)			
Alone	26.1	30.2	27.5
With spouse	38.6	37.3	41.6
With others	30.6	27.4	25.1
Group quarters	4.7	5.1	5.8
Race (%)			
Nonwhite	12.9	10.5	10.1
White	87.1	89.5	89.9
Economic status			
Monthly median family income[a]	$1,005	$1085	$1025
Aged below national poverty threshold (1979) (%)	17.0	13.9	14.8
Aged receiving SSI (1980) (%)	7.3	7.4	7.2
Transfer program involvement			
Aged enrolled in Medicare (1981) (%)	97.3	97.1	97.2
Medicare expenditures per aged resident (1981)	$102	$119	$109
Aged Receiving Medicaid (1980) (%)	13.9	12.3	13.4
Medicaid expenditures per aged resident (1980)	$22	$37	$28

[a] In 1986 dollars.

The set of relevant community descriptors included a description of long-term care and related services in the demonstration areas. It was important to describe these long-term care services for three reasons: first, they were the service resources available to channeling case managers for care planning; second, they represented what was available in the absence of channeling to control group members; and third, documentation of these systems may be helpful in interpreting treatment–control differences in outcomes. These long-term care and related services included:

- Skilled nursing services at channeling sites.
- Receipt of skilled and other in-home services.

- Homemaker services with personal care.
- Receipt of transportation and home-delivered meals.
- Licensed nursing home facilities.
- Reported waiting time for nursing home placement.
- Characteristics of short-stay hospital bed supply.
- Physicians engaged in patient care activity and physician shortage areas by site.
- Availability of community long-term care services.
- Availability of nursing home beds.

In summary, it is important to be sensitive to the influence that relevant community characteristics have on a program. They represent potentially significant external influences that may affect the program's impacts. But they also affect the generalizability of your program and thus should be included in the process analysis.

C. Family Characteristics

There is a large body of evidence indicating that family involvement, support, and agreement with the goals of a program are significant influences on the program and its outcomes. For example, in a survey associated with the LTC project, it was found that 89% of sample members had one informal care giver and 60% had two or more. Spouses and children were found to play the major roles in informal care giving, but other relatives and nonrelatives also provided some care in about half of the care-giving networks. Similarly, family involvement has been shown to be a correlate of successful community placement for developmentally disabled adults (Schalock, Harper, & Genung, 1981), assessed quality of life indices (Schalock & Lilley, 1986), and postsecondary community placement success of handicapped students (Schalock *et al.*, 1986). Thus, don't overlook the role that family members, volunteers, benefactors, and friends play in impacting your program and its outcomes.

The list of family characteristics potentially impacting your program can be extensive, so keep the mother-in-law test and our 20th guideline clearly in mind. Our best advice is to talk to your participants and your staff, and ask them to generate a list of persons outside program personnel who provide help or services for the participants. The list might surprise you. Once you generate the list, you might want to develop a simple quantification procedure to measure the amount of involvement. Representative scoring categories include amount of time, number of services provided, or a simple degree-of-support or involvement scale. These measures and their results can then be added to your process analysis and provide valuable input to both you and your readers. We realize that there is no way to capture all the family characteristics that might impact your program; but it is important to at least be sensitive to the more critical ones and attempt to quantify their magnitude.

V. Summary

In this chapter we focused on process analysis from the program's perspective and looked specifically at three groups of factors that influence a program and its outcomes. The first group relates to the organization's structure and includes the organization's legal designation, governance structure, funding considerations, and major service components. The second group is composed of three internal factors—the program's philosophy and goals, phase of program development, and resources. The third group includes three external factors involving formal linkages, community descriptors, and family characteristics. Throughout the chapter, we provided examples of how these factors might impact a program and why they are important to include in a process analysis.

We also presented three guidelines that should help you decide on what type of data—and how much of it—to include in a process analysis from the program's perspective. The three guidelines are:

- Describe your program and its environment in enough detail to allow its replication.
- In reference to internal factors, never assume that what staff are hired to do is what they actually do.
- Consider only those external factors the environmental effects of which exceed some threshold and thereby can reasonably be expected to affect the program and its outcomes.

In this chapter, we covered the second of three components of a process analysis. Chapter 4 focused on the persons served and the services provided by the program. Chapter 5 has discussed how those services are provided within the context of the program's organizational structure and environment. Chapter 6 addresses costs and costing procedures.

VI. Additional Readings

Beer, M. (1980). *Organization change and development: A systems review.* Santa Monica, CA: Goodyear Publishing.

Posavac, E. J., & Carey, R. G. (1980). *Program evaluation: Methods and case studies.* Englewood Cliffs, NJ: Prentice-Hall.

Rutman, L. (1977). Formative research and program evaluability. In L. Rutman (Ed.), *Evaluation research methods: A basic guide* (pp. 59–71). Beverly Hills, CA: Sage.

Stake, R. F. (1986). *Quieting reform: Social service and social action in an urban youth program.* Champaign: University of Illinois.

Washington, R. O. (1980). *Program evaluation in the human services.* Washington, DC: University Press of America.

Wildavsky, A., & Tenebaum, E. (1981). *The politics of mistrust: Estimating American oil and gas resources.* Beverly Hills, CA: Sage.

6

Analysis of Program Costs

I. Overview

After describing who is served and the services that are provided, the next most important evaluation task is to determine what the services cost. Being able to account for the resources used by a program is a crucial responsibility of any program management. It is also important for describing the intensity of the services provided, for budgeting program replications, and for evaluating whether the impacts produced by the program were sufficiently large to justify the program.

Of all these reasons, accountability is the most familiar. Funders, administrators, and consumers all have an interest in knowing whether the money allocated to a program is being spent for its intended purpose. A first step in addressing this question is to determine exactly how much money was spent and what it was used to purchase. This is the goal of most program financial accounting systems and is a necessity for all programs.

The analysis of program costs is also an important part of the project description. It provides a quantitative measure of the resources used to operate the program. Such information helps to indicate the *intensity* of the intervention (a $200 per client program is different than a $2000 per client program) and is essential background information for anyone interested in budgeting a program replication.

Finally, cost analysis provides a foundation for interpreting the findings of an impact analysis. To a large extent, the magnitude of the effects produced by a program will be judged by the resources used to produce them. This comparison of program effects and resource use is called benefit–cost analysis and has proved to be a useful tool in assessing the performance and desirability of human service programs (see Chapter 9).

For all of these reasons, persons administering or evaluating human service programs (or any programs for that matter) should be interested in costs and cost analysis. Most programs have, or should have, an accounting system that that will provide the basis for the analysis, and the basic elements of a cost analysis can be produced from this accounting system without much effort. These basics will be essential to anyone who wants to act on a program's evaluation information, since it is impossible to begin to implement a program or policy without understanding the costs involved. Program evaluations that fail to present cost estimates lack an essential reference point for implementation and will have little value to persons who want to translate those evaluation findings into action.

Our approach emphasizes that cost analysis can proceed in increments. All programs should seek to provide some basic cost information about their activities, information that can be culled directly from their accounting systems with little work. Programs can then choose to examine more complex issues, all of which build on the basic estimates of cost. These additional issues provide valuable information but require more work and more data to analyze. The extent to which it is useful to pursue these additional issues will depend on the developmental stage of the program, the interests of the analyst, and the information demands of policymakers.

The chapter is divided into two sections dealing with (1) our approach to cost analysis and (2) estimating costs. Contrary to the two previous chapters, we do not present data collection forms or table shells, since we feel that those activities are best done in reference to the benefit–cost analysis procedures we outline in Chapters 9 and 10. In reading the chapter, you may want to refer to Table 6.1, which outlines the chapter and also presents a number of general guidelines regarding the analysis of program costs.

II. Approach to Cost Analysis

Cost analysis often seems like a world of arcane rules. Cost analysts worry about expenditures, resource costs, interest rates, capital cost, amortizations, accruals, present values, and inflation adjustments. While these concepts are important, they should not intimidate persons from using cost analysis either as consumers or producers. The rules follow from some fairly straightforward principles, and most users need know only those general principles rather than all of the resulting specific rules.

Accounting rules essentially are intended to help ensure that all costs of a program are captured in the analysis and that costs are consistently measured. In this regard, the analyst must carefully define the program being studied and then proceed to identify the costs incurred to operate that program. Virtually all of the analytical work in the cost analysis is in developing a clear and precise definition of the program. Thus, time spent determining (1) exactly what is being studied, (2) the procedures for valuing the resources used by a program, and (3) the analytical perspectives to be used will pay off many times by simplifying the tasks of estimating costs.

Table 6.1. **General Guidelines regarding the Analysis of Program Costs**

A. Importance of determining program costs
   1. Accountability
   2. Program description
   3. Benefit–cost analysis
B. Approach to cost analysis
   1. Describe the program
      a. Follow the program description established in the set-up
      b. List all the resources used by the program
   2. Value the resources used
      a. Value(s) based on opportunity costs
      b. Measured by market price
   3. Perspectives on costs include program, government, and participants
C. Estimating costs
   1. Total costs to the program
      a. Define the period over which costs will be measured
      b. Use actual recorded costs
      c. Cost analysis period should correspond to the impact analysis period
   2. Average costs to the program
      a. Advantages to using average costs
         (1) Easily compared with estimates of average program effects
         (2) Allows focusing attention on issue of service intensity
      b. Calculation methods include average cost per person or per enrollee
   3. Estimate costs to government and participants

A. Describing the Program

Like most of the other aspects of program evaluation, the cost analysis begins with the program definition that was specified in the setup. The cost analyst uses this definition to develop a specific view of the program activities. As strange as it may seem, this view does not begin with the program accounting system or with dollars and cents. It begins with a view of the use of resources by the program; that is, it begins by looking at the activities required to produce the program services, including the various staff members, the facilities, any materials and supplies, the time, other resources provided by the persons served by the program, and any services or materials provided by other agencies. It is only after these resources have been carefully identified that the cost analysis turns to the issues of dollars and cents.

In describing these resources, it is essential to follow the program definition established in the setup. Otherwise, persons using the evaluation may be misled about the actual costs of providing the program services described in the process analysis or the costs of obtaining the effects estimated in the impact analysis. The analyst should regularly consult the definition included in the setup to see that the cost analysis is proceeding consistently. If there is a difference, it should be resolved so that reported costs match the reported program services and effects.

The focus on resource use—that is, on the activities rather than on just the flow of dollars—is crucial to being able to use the cost analysis as a means of describing the intervention. It is also essential for persons who want to understand what happened in the program and who want to use that information to budget or modify their own efforts.

Often this focus on the program activities requires the analyst to go beyond the financial books of a program. The analyst must examine donated goods and services, off-budget costs, and other instances in which the accounting records may provide an incomplete or misleading picture of the level of activity needed to operate the program. It is only by looking at this underlying level of activity that the analyst can understand the program and its costs.

To see an important implication of the focus on resource use, consider a case in which the program being evaluated is operated by a larger agency that runs many other programs. This larger, or "host," agency may provide many resources to the program. These include the use of agency facilities, bookkeeping or accounting support, and the advice and expertise of the agency's senior management. The cost analyst must identify these types of support and include them in the analysis. Otherwise, there is the risk that persons who try to replicate the program will unintentionally underbudget their efforts, a situation that will be embarrassing, at least, and could have more serious repercussions if the excluded cost items were crucial components of the program.

The rule to follow in a cost analysis is to list all resources used by the program, even if they cannot all be valued in dollars. For the most part, this list will come from the setup. The listing of all resources will help consumers of the analysis make better judgments about whether the evaluation results can be applied to other situations. Consumers need to see all of the resources required by the program so that they can determine how best to supply those resources to any new program efforts. In this way, the cost analyst is more concerned with the total resource costs of the program than with the specific funding and support arrangements made in the program under study.

Another example of why it is important to consider resource use, and any special circumstances surrounding a program, can be taken from efforts to budget programs based on model programs operated in university settings. Programs operated at a university can often draw on a variety of special resources and supports that may be unavailable to other programs. For example, university programs may use students to staff the program. Students represent a special labor market, since they are often willing to accept lower wages in exchange for the opportunity to learn from the program. Students may also differ from the workers who would staff a program fielded in a different environment in terms of their demands for full-time hours, shift pay differentials and fringe benefits. The university-sponsored program may also receive subsidies from the university in the form of free advice from faculty, free (or below-cost) use of university facilities, and accounting assistance. Any person attempting to replicate a university-based program will have to consider whether a new program will have similar access to these resources, and the only way that it is possible to consider this important issue is if the cost analyst took the effort to describe resource use rather than just examining direct program expenditures.

B. Valuing Resource Use

After the resources used to operate the program have been identified, there is the task of determining their value. The key concept to use in valuing these resources is that of *opportunity cost*. This is also the concept that is used to value program effects in the benefit–cost analysis presented in Chapter 9. Opportunity cost, like many of the other elements in an evaluation, is based on comparisons. In this case, it is the comparison between the use of the resource and what that resource would have been used for in its next best alternative. Specifically, the opportunity cost of any action or thing is what must be given up in order to take that action or obtain that thing. This notion, like our discussion about listing the program costs, focuses on resources and their alternative uses.

While this concept of cost may initially seem particularly arcane, it is, in fact, a concept used most of the time. We often talk about the cost of doing one thing in terms of what we had to give up in order to do it. For example, a parent may remark that a parent–teacher meeting cost an evening at the ballgame, or a person who must go to court because of a traffic ticket may describe the cost of the ticket as the fine plus the day of work lost while the person was in court. In the same way, the cost of reading this book is whatever you would have otherwise done with the time—read another book, complete a specific task for work or school, do some shopping, or even daydream. For a program administrator, the cost of funding a specific program is the forfeited alternative use for the resources used in the program.

For the most part, the best alternative use of a resource is measured by its *market price*. This is because the workings of a competitive market, in which producers and consumers continually interact, lead to a situation in which the prices of goods and services accurately reflect their value, both in terms of their value to consumers and their costs of production. Typically, there are many consumers willing to pay the market price of a good, thereby indicating that the value of the good is worth at least that much to them. Thus, the opportunity cost of the good—its value in its most likely alternative use—is measured by the price of the good in the market. It is this feature of the marketplace that has so endeared it to economists.

To the cost analyst, this feature means that the costs recorded in the program accounting system will usually indicate the true value of the resources used by the program. For example, the wages paid to program staff are in line with what those persons could earn in alternative employment, and the prices paid for facilities, equipment, and supplies indicate the values those goods have in the market.

The major instances in which this is not true pertain to goods and services that are donated to the program—for example, volunteer labor, office space provided at below market prices, or material given to the program—and resources used by the program but paid for by other agencies—for example, counseling or medical assistance provided by another program to participants in a training program. The analyst needs to identify the various resources used by the program and trace them back to the person or agency that provided them.

C. Perspectives

The third key principle in cost analysis is to specify the perspective or perspectives from which costs will be measured. This is important, because the costs of a program will be viewed differently by different groups in society, as seen in the example we present in Chapter 9 (Table 9.7).

When people think of program costs, they generally think of the expenditures made by the program. These include the wages and fringe benefits paid to staff and the costs of the materials and supplies used by the program. However, there are other perspectives of the program that may perceive costs differently. For example, if the program leads participants to increase their use of other government services, then when the program is viewed from the perspective of the government as a whole, total costs will include the direct program expenditures as well as the increased expenditures incurred by other government programs.

In the same way, participants in a program will view costs differently than will the program administrator. Participants will not be concerned directly with the program budget or indirect effects the program may have on the budgets of other programs; instead, they will focus on any fees they must pay and the amount of time they must spend in the program. The time or, more precisely, the opportunities forfeited while they spend time in the program often constitutes the major cost of participation for many participants. For example, students in college will incur direct costs in the form of expenditures for tuition, books, and other fees. They will also incur a cost in the form of the earnings they would have received if they had taken a full-time job rather than enroll in college.

In sorting out the perspectives, it is useful to start small and build up. Add in only as much complexity as you need to answer your evaluation questions. As noted earlier, the key is to follow the resources used by the program and trace those resources back to whoever pays for them. Most analysts start with direct program costs and then move to other perspectives as needed to address the evaluation goals. In particular, the cost analyst must consider whether the benefit–cost analysis needs information about specific perspectives or whether a particular funding source needs information about the net effect the program has on the aggregate expenditures of that funding source. Thus, an important guideline:

Guideline 21. When considering the costs of a program, always start with the program definition established in the setup and list all resources used by the program, regardless of who pays for them. Also, indicate any circumstances in which the program may have access to special resources or pays prices that are different from those prevailing in the regular marketplace.

III. *Estimating Costs*

The cost analysis begins with the estimation of the total costs incurred directly by the program being studied. Because estimates of total cost are often difficult to compare with the estimates of program impacts or with the costs of other programs, the analyst usually divides the total cost estimate by the number of persons served to calculate average cost. From this starting point, the analysis can be expanded to encompass other perspectives. Here, we consider two perspectives other than that of the program: the perspective of the total government budget and that of the persons who participate in the program.

A. *Total Costs to the Program*

The program budget perspective, being easiest to measure, is the place to start. A good accounting system will generally provide adequate information to estimate the total costs of the program as seen from the perspective of its own budget. The principal areas in which care is needed are in the treatment of capital expenditures and changes in accrued costs.

In order to make the analysis manageable, it is first necessary to define the period over which costs will be measured. This is particularly true for an ongoing program that may have been incurring costs for many years. It is generally best to pick a recent period for which complete accounting records are available, usually the most recently completed program fiscal year. By picking a period for which the books are closed, you can avoid the problems of estimating what final costs will be or changing your analysis when the final numbers become available.

It is also important to use actual recorded costs rather than basing the cost analysis on budget or planning documents. There is essentially nothing to study until actual cost information becomes available. Before that time, a cost analysis would be merely reflecting the assumptions and estimates used to develop the budget rather than providing information about the resources actually needed to implement the program or policy.

If you plan to conduct a benefit–cost analysis, or otherwise link the cost analysis to the impact analysis, then it is also important that the period used for the cost analysis corresponds to the period used in the impact analysis. This is done to ensure that costs and benefits correspond to the same intervention and thus are directly comparable. If you plan to study the program effects on a specific group of persons, then the costs from the period when those persons were served should be the ones included in the cost analysis.

Once the period has been defined, the analyst totals up the costs recorded for that period. This is relatively straightforward, except for a few instances in which the costs recorded in any given year may not accurately reflect the actual resources used by the program in that year.

The most common instance in which expenditures in a year do not reflect actual resource use is when the program makes capital purchases. Capital expenditures are generally defined as expenditures for items that can be expected to last for more than 1 year. To simplify the accounting procedures, there is usually a dollar minimum (generally $500) specified for capital expenditures so that only items costing more than that amount will be included as capital purchases. These purchases typically include office furnishings and equipment, buildings, cars or buses, and specialized equipment needed to provide the program services.

The problem these capital expenditures pose for the cost analysis is that they provide a stream of services often lasting over several years. Thus, when estimating the costs of a program for any time period shorter than the life of the capital goods, the analyst wants to include only the value of the services used in that period, not the value of future or past services. This situation can be seen, for example, in the case of a building purchased by a program. The purchase of the building requires a large initial expenditure (generally financed with a mortgage or other loan) but provides service in the form of space for many years. In this case, the cost analysis should include only the value of the building's services that were used during the period covered by the cost analysis.

To measure these costs, there are two alternative procedures. The first is to estimate the value of the services by calculating the cost of renting the building (or other capital item) for the period covered by the cost analysis. These rent estimates can typically be obtained from rental agencies or real estate firms. The other method is to calculate how much of the building was ''used up'' during the period by estimating depreciation. In either case, any ongoing costs associated with the building (maintenance and utilities, for example) would be included, since they represent resources consumed during the period of the analysis.

The other major area in which expenditures during a period may not reflect actual resource use is when the program accounts are maintained on a cash rather than accrual basis. The essential difference between these two systems is that the cash system monitors only the net flow of cash in and out of the program. Under this system, the accounts would exclude the costs for those resources the program used in one period but did not have to pay for until a later period. Similarly, cash accounts would include expenditures made during one period in order to pay for resources used in a prior period. Under an accrual system, the accountant tries to include in the books for any period the costs for the activities that took place during that period. Clearly, for an analyst interested in measuring the value of resource use in a particular period, the accrual system offers substantial advantages.

The two systems will differ when an agency prepays some of its bills, when it receives grants for services to be performed in the future, or when it does work for which it will be paid in the future. In all of these cases, the cash flow in a period will differ from the actual use of resources.

In an ongoing program, revenues and expenditures will generally balance each other over a fiscal year. This is particularly true in human service programs that operate with a relatively fixed amount of working capital and so cannot afford to prepay many bills or defer receiving payments to a great extent. Thus, even if a program does keep its books on a cash basis, if there is no change in the amounts of (1) payments for past activities or (2) deferrals of payments for current activities into future periods, then the analyst can use the expenditures recorded for the period as a measure of the activity that took place in that period. In comparison, if a program is growing or otherwise changing, then it may be accruing costs differently than indicated in the cash-based accounts. In such an instance, the analyst must subtract out any expenditures made during the period to pay for resources used in a previous period and add in the costs for any activities where actual cash payments have been deferred.

In estimating total costs, it is important to keep in mind the level of precision needed from the analysis. It is often unnecessary to resolve all the issues about capital costs, accruals, or other cost factors. In many cases, it is clear that such refinements will only make trivial differences in the cost estimates. When that is the case, it is usually most efficient to estimate the general magnitude of the costs.

One way to determine whether a cost issue will result in a trivial change is to assess the maximum effect it could have on the estimate of average costs. If this maximum effect would change the average cost estimate by less than 5%, then it will usually be safe to ignore that particular cost issue. The exception to this rule would be if precision was very important to the analysis, in which case a more precise standard could be used for determining which cost issues to resolve. For the most part, the 5% standard is adequate and easily encompasses the range of accuracy found in most cost studies of human service programs.

B. Average Costs to the Program

While we initially focus on total costs, they are often only an intermediate goal of the analysis. Much of the analysis, particularly comparisons of costs and impacts and the comparison of costs across programs, is done using average costs. Average costs are more easily compared with estimates of program effects that will generally reflect the effect the program has on an average enrollee (see Chapter 9). Average costs also offer an important advantage when looking at the costs of several programs. Total costs are determined to a great extent by the number of persons served; larger programs will generally have larger costs than comparable smaller programs. This means that total costs reflect program scale as much as the intensity of services provided to the typical participant. By looking at the average costs of a program, the cost analyst can abstract from the scale of a program and focus attention on the issue of service intensity.

Average cost is generally calculated by dividing the estimate of total cost for a specific accounting period by the number of persons served during that period. In making this calculation, it is absolutely essential that the estimates of total cost and the number of persons served by consistent. The total cost should include all costs of serving the persons included in the figures on the number of persons served, and only the costs of serving those persons. Otherwise, the estimate of average cost will be misleading. It will be too low if costs are excluded or too many persons are included in the estimate of persons served. Average cost will be too high if the costs of serving persons not included in the denominator are included in the numerator.

Figure 6.1 illustrates the types of problems that face a cost analyst trying to

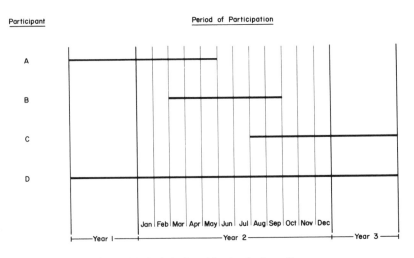

Figure 6.1. Period of participation for four clients.

estimate average costs. In the figure, the accounting period for estimating total costs is indicated as Year 2, and the lengths of participation for the four clients in the program are indicated by the lines marked A, B, C, and D. If the analyst divides total costs for Year 2 by the number of persons served, average costs will be underestimated because some of the costs of serving clients A, C, and D will be excluded. In particular, the costs incurred to serve these persons during Years 1 and 3 will be left out of the estimate based only on costs incurred in Year 2.

This issue of insuring that the cost data and participant data pertain to the same time period and the same individuals must be dealt with correctly in your cost estimation, for otherwise it will significantly throw off all subsequent analyses. You can deal with the problem by using either (1) average costs per participant month of service or (2) average costs per enrollee.

   a. Average Costs per Participant Month of Service. This cost can be estimated by using the following formula:

$$
\begin{bmatrix} \text{Average cost} \\ \text{per participant} \end{bmatrix} = \begin{bmatrix} \dfrac{\text{Total costs for period}}{\begin{array}{c}\text{Total months of participation} \\ \text{for period}\end{array}} \end{bmatrix} \times \begin{bmatrix} \text{Average months of} \\ \text{participation per} \\ \text{participant} \end{bmatrix}
$$

This formula requires that you:

1. Determine your total costs for the specified period as discussed earlier in this section.
2. Determine the total months of service (one client served for one month equals a month of service) for the clients served during the months encompassing the time period for which you want costs. In reference to Figure 6.1, the months of service in Year 2 would be 28.
3. Divide the total costs by the months of service, which will give you the average cost per month of service.
4. Multiply the product of step 3 by the average length of stay (average number of months per client), which gives you the average cost per participant.

The advantage of using this approach to determining average costs to the program is that it ensures that cost data and participant data pertain to the same time period and the same individuals. Thus, the measure will permit valid cross-program comparisons.

   b. Average Costs per Enrollee. If one is involved with a steady state program, then rather than determining the average cost per month of service one can determine average costs for those enrolled only in a particular period of study, such as Year 2 in Figure 6.1. This procedure does require a number of assumptions which are illustrated in Figure 6.1. One assumption is that the costs of persons enrolled in Year 2, but served in Year 3 equal the costs of those enrolled in Year 1 and served in Year 2. Second, one assumes the same number of "A" clients as "C" clients. The duration of service is the same for those two groups, even though the time distribution of service is different. Thus, the total costs of providing service in Year 1 to all the type-A and type-D clients must equal the costs of serving type-C and type-D clients in Year 3.

If these assumptions can be met, then the average cost per enrollee is estimated with the following formula:

$$\frac{\text{Average cost}}{\text{per participant}} = \frac{\text{Total cost for time period}}{\substack{\text{Number of people enrolled} \\ \text{during that period}}}$$

If the assumptions cannot be met, such as would be the case if your program has a large number of type-D clients or if the program is growing so that there are more type-C clients than type-A clients, then you will need to use the first method (average costs per participant month of service).

Because of the importance of insuring that the cost data and participant data pertain to the same time period and the same individuals, we offer the following guideline:

Guideline 22. When computing average costs, be absolutely sure that the total costs in the numerator are measured consistently with the number of participants in the denominator.

C. Costs to Government

Estimating costs involves tracing all resource use for the program in question. For example, suppose that you are estimating costs for a community-based residential program that uses a workshop or activity program during the day. It would be incorrect to look only at the residential costs, for you also need to include the costs to government of the workshop or day activity program. As you might remember, we discussed this issue in Chapter 3, in reference to the longitudinal study of the court-ordered deinstitutionalization of Pennhurst residents (Ashbaugh & Allard, 1983). The initial analysis suggested that the costs to maintain persons in Pennhurst were significantly greater than the costs for community placement. On closer analysis, however, when the costs of providing day programs in the community were factored in, the estimated costs of the two programs were essentially the same.

Most human service programs use governmental resources that frequently do not appear in the program's financial accounts. These include surplus goods (for which the program might pay only transportation charges), meal costs supplemented by food stamps or reimbursed by the National School Lunch Program, and medical services provided by state and local agencies. Thus, it is essential to trace all resources used by the program in question. If one looks only at the direct costs of a residential program, for example, one cannot replicate it unless all costs are factored into the total and average program costs. Therefore, as either an evaluation producer or consumer, one needs to focus on the broader perspective on costs. We will do this in Chapter 9 on benefit–cost analysis.

D. Costs to Participants

Persons who participate in human service programs forgo other opportunities. They may have to pay fees, delay getting a job, give up one job for another, or spend time in an education or training program that precludes other potentially beneficial activities. These costs, referred to as opportunity costs, should be included in one's cost estimates. Youths who participated in the Job Corps (Thornton *et al.,* 1982), for example, had to forgo employment opportunities they otherwise would have taken. Thus, the wages they would have earned are a cost to them for participating in the Job Corps. College students forgo the same type of opportunity costs. We also return to this issue and discuss it in considerable detail in Chapter 9 on benefit–cost analysis.

IV. Summary

Program analysis frequently involves the analysis of program costs. As Demone and Harshbarger (1973) state, "recommendations for programs whose costs are not estimated are simply statements of philosophy that will not be taken very seriously" (p. 9). As is evident from your reading, this was not a chapter on cost accounting, since we assume that your agency already has an established accounting system. Rather, what we have tried to do in this chapter is present a conceptual approach to the analysis of program costs that results in the accounting of the resources used by the program. Such an accounting is essential for a number of reasons, including accountability, program description and replication, and providing the foundation for interpreting the findings of an impact analysis.

We organized the chapter around our approach to cost analysis and estimating costs. Our approach to cost analysis has three components, namely, describing the program, valuing resource use, and different perspectives on cost. Estimating costs involves determining total costs to the program, average costs to the program, costs to the government, and costs to the participants. We suggested the following two important guidelines regarding the analysis of program costs:

- When considering the costs of a program, always start with the program definition established in the setup and list all resources used by the program, regardless of who pays for them. Also, indicate any circumstances in which the program may have access to special resources or pays prices that are different from those prevailing in the regular marketplace.
- When computing mean costs, be absolutely sure that the total costs in the numerator are measured consistently with the number of participants in the denominator.

And finally, we emphasized that cost analysis should probably proceed in increments. All programs should provide some basic cost information about their activities. Thereafter, programs can choose to examine more complex issues that build on the basic estimates of cost. These additional issues provide valuable information, but they require more work and data to analyze. The extent to which it is useful to pursue these additional issues will depend on the developmental stage of the program, the interests of the analyst, and the information demands of policymakers. Each of these issues is pursued in greater depth in subsequent sections of the book on impact and benefit–cost analyses.

V. Additional Readings

Horngren, C. T. (1982). *Cost accounting: A managerial emphasis* (5th ed.). Englewood Cliffs, NJ: Prentice-Hall.

Killaugh, L. N., & Leininger, W. E. (1987). *Cost accounting: Concepts and techniques for management* (2nd ed.). New York: West Publishing.

Levin, H. M. (1983). *Cost-effectiveness: A primer.* Beverly Hills, CA: Sage.

Moriarity, S., & Allen, C. P. (1987). *Cost accounting* (2nd ed.). New York: Harper & Row.

III

Impact Analysis

Impact analysis focuses on a program's effects or impacts on the targeted population. It is an essential component of a program evaluation, but it can be quite difficult to accomplish because it involves making comparisons between what happens in the program and what happens in the comparison state. In impact analysis, therefore, you need to make statements about both sides of the comparison. Program administrators frequently do not look at the comparison group—those that did not enter the program—or stop and ask, what would have happened to my participants had they not entered the program? Impact analysis involves data collection, following people over time, and thinking about what actually happened to the participants and what would have happened had they not been served. Thus, it involves marshalling the evidence and interpreting the findings (see Table 3.1).

Impact analysis addresses a number of critical evaluation questions that are the basis for benefit–cost analysis and gives answers to questions needed to make policy and other decisions. Exemplary questions include:

- Did the program have the intended effects on outcomes?
- How big are the effects?
- How much uncertainty is there surrounding the estimate of each effect?
- Can these effects be attributed with reasonable certainty to the intervention being studied?

In this section of the book we review techniques to assist administrators understand impact evaluation literature. There are lots of important lessons to learn from that literature, but many of the statistical and procedural techniques used are not within the scope of either the book or the typical administrator. We present approaches that can be used if you are an evaluation producer, but we caution you that the suggested procedures result in estimates with substantial uncertainty. Thus, we use a "broad brush" approach to outline some specific, but rough, impact analysis techniques that can be done with relative ease by an administrator and that can be used to assess your program's potential in meeting those goals and objectives we discussed in Chapter 3. But again, we caution against pushing too far; know your limits and stay within them.

The section is divided into two chapters—one on measuring program outcomes and one on estimating program impacts. In thinking about outcome measures, go back to the setup in Chapter 3 and determine whether the measures selected really capture the important elements of the comparison. For example, if the program is to impact employment, are the measures capturing the *impact* of employment, or merely *placement* into employment? Impact analysis takes a long time to develop, since the impacts from employment, such as changes in economic and social conditions, also take a long time to develop. Thus, if you do not have the luxury of time to follow participants, you will need to focus on *performance indicators,* which represent short-term outcomes that are clearly tied to or strongly predict the impacts of interest. For example, placement rates are common performance indicators used in employment programs. They first appeared in the evaluation literature after World War II, when veterans were being trained and placed quickly into new jobs. But placement rates are inappropriate to use if you want to demonstrate the long-term effects or impacts of an employment program. In that situation, you might want to focus on the effects on labor market performance, education and training, transfer program use (such as SSI), and social and living conditions.

In reference to schools, for example, one could look at grades and graduation as outcomes; but if the goal of education is life-style changes, additional indicators would be needed to determine the impact of the education received.

Our intent in this section is also to assist you as the consumer in evaluating whether the measures used in the evaluation you are reading capture the important elements of the comparison. For the producer, we use a broad brush approach and suggest some methods and potential comparisons that will allow the administrator to better handle the program and convey the program's outcomes and potential impacts. Our advice to the producer is that if you have short-term data (such as placement rates), treat these as performance indicators, and don't imply that they are impacts. If, however, you have longitudinal data on your program, participants, and outcomes, then go ahead and set up some comparisons that will get you further into an impact analysis. But we again stress that you are not doing a complete impact analysis. Hence, the following caution:

Caution. The techniques we present result in only a partial impact analysis. Impact analysis involves certainty and precision; what you will have from our techniques is a rough idea.

The above caution should not preclude considering impact analysis, although our general feeling is that most producers will benefit more from conducting process analysis. If, however, you are concerned about your program's outcomes and potential impacts, then the two chapters will be well worth your time. Impact analysis does require precision as related to experimental design, sample size, data collection procedures, and analysis techniques. If you have a small sample size and limited, short-term outcomes, then you really can't do impact analysis. It's like our fish analogy in the Preface: don't bring a block and tackle to catch a bass. But you're still a consumer, and the techniques we present should help you significantly in that role. Additionally, just the process of going through the two chapters will help you better understand your program and others'.

This section on impact analysis contains two chapters. Chapter 7 deals with measuring program outcomes. Subsections include suggestions regarding what outcomes to measure, when to measure them, how to measure them, and where to measure them. Throughout the chapter, examples will be taken from the Structured Training and Employment Transitional Services (STETS) demonstration project that was first introduced in Chapter 1. As you might remember, the STETS project attempted to determine the effectiveness of transitional employment programs in integrating mentally retarded young adults into the economic and social mainstream of society. The chapter also contains a number of guidelines and cautions regarding the measurement and meaning of _program outcomes_ that reflect the program's activity.

Chapter 8 deals with estimating _program impacts,_ which is not easy since one needs to measure both sides of the structured comparison. Estimating the program's impacts involves comparing the outcome measures for a group of participants with those for a comparison group. Because of the complexity of impact analysis, and the field guide nature of this book, we focus primarily on the evaluation consumer, although we do discuss two practical impact analysis techniques that most producers can use. The chapter has four sections, in which we discuss techniques one can use to estimate the effects of a program, a proposed impact analysis model, two practical impact analysis techniques, and three validity issues surrounding impact analysis.

Throughout the section on impact analysis, we caution the reader about the effects of assumptions and methodologies on the certainty and precision of our impact estimates. Impact analysis is tough, and it requires rigorous techniques. But one should not shy away from recognizing its importance in and of itself, as well as for providing the basis for benefit–cost analysis. The latter cannot be done without estimated or programmatic impacts.

Measuring Program Outcomes

I. Overview

Administrators are faced daily with questions from various constituents regarding how well their program is doing and whether it has really had an effect on the participants. Answering these questions requires clearly defined and measured outcomes. In many programs, the time hasn't been taken to implement the necessary steps to have relevant outcome measures on participants, despite their importance. Part of the reason is the focus on process analysis; part is due to ignorance of the literature regarding currently accepted outcome measures; and part of the reason is simply lack of time to get the data on outcomes or to develop the techniques to capture outcome measures. Our general feeling is that administrators *can* operationalize outcome measures, but they simply haven't, due to financial, time, or "perpetual crisis" restraints. Thus, our attempt in this chapter is to provide you with a "broad brush" approach to selecting and measuring appropriate outcome measures that can be used for a number of purposes, including reporting, program evaluation, and as the basis for impact and benefit–cost analyses.

Outcomes are essential to measure because they reflect *what* we are trying to accomplish in human service programs. The process, and the analytic techniques we discussed in the last chapter, is the *way* we accomplish outcomes. Think of a flight to Los Angeles, for example. The important thing is what you do when you get there, not the flight. If, for some reason, you end up sitting on the tarmac and have fun doing it, that's probably okay *if* your goal was only to have fun. But possibly you wanted to get to L.A. for other reasons. Either way the goal is getting there, and that is the outcome. So in this chapter think *outcomes,* not process.

There are a number of advantages to including outcome measures in your planning and service delivery efforts. For one thing, they allow you to see if you are accomplishing what you wanted to accomplish. This is also what people to whom you are accountable will want to see. Second, in regard to proposal writing, a good setup and rationale that links specific processes to desired outcomes will communicate to potential funders that this person is truly in control, which should increase your funding chances. It's another way you can show others that you have control of your program and can communicate that to the outside world, including passing the mother-in-law test.

In the chapter, we present suggestions regarding what outcomes to measure, and we outline techniques regarding how to measure the outcomes you've selected. We also present a number of guidelines. Generally speaking, we suggest that you stick with objective outcomes that can be measured within the time period you have, and that the resources devoted to the process are appropriate to what you think you will get out of it. Thus, don't go overboard in measuring outcomes, for you still have to ensure that your program runs. Be sure, in other words, that your resources are balanced. Our approach is broad brush, and we remind you that one cannot paint small distinctions or objects with a broad brush.

II. What to Measure

What one measures in impact analysis depends upon the program's goals and objectives. Think back to Chapter 3, the setup and the rationale that specified the link among program participants, services received, and expected outcomes. As a consumer or a producer, you need to ask, do these measures result logically from the program; and furthermore, do the measures capture important elements of comparison on both sides (program participants and nonparticipants)? For example, if you administer an employment training program, then the outcomes should relate to labor market performance areas, such as income, that can be used to compare participants with nonparticipants. Therefore, as either a producer or consumer, you should tie outcome measures to the program's goals and objectives, keeping in mind the comparison you are making. This leads to our first guideline in this chapter:

Guideline 23. Outcome measures should be tied to a program's goals and objectives; keep in mind the comparisons you are making.

Before considering specific outcome measures, we would like to suggest five criteria that are helpful in determining what to measure. They include using measures that (1) can be attributed with reasonable certainty to the outcomes of the program, (2) are sensitive to change and intervention, (3) are obtainable, (4) are objectively defined, and (5) are prioritized in regard to how they can be used to proxy all the other potential outcomes.

Attributed to program. One should be able to see clearly that there is a link between these people getting these services and these outcomes. The Structured

Training and Employment Transitional Services (STETS) research plan, for example, was designed to address the following questions:

- Does STETS improve the labor market performance of its participants?
- Does STETS affect the use of alternative programs by participants?
- Does STETS participation help individuals lead a more normal life-style?

The outcome measures used are those presented in the top four categories of Table 7.1. As a job-training program, one would reasonably expect that it would increase employment levels and improve the quality of jobs held by program participants. Similarly, if its intent was to move individuals out of training programs and schools into competitive employment, it thereby should reduce the long-term use of training and education programs. The employment outcomes should also have secondary effects on public transfer dependence, since the receipt and amount of most types of transfers are income conditioned. And finally, it might also be expected that STETS would have impacts on other areas of participants' lives—especially their overall economic status, their independence in financial management and living arrangement, their use of formal and informal services, and their general level of involvement in regular, productive activities. However, because of resource constraints, these measures were not pursued as thoroughly as the employment and transfer data sets. The demonstration did, however, collect some data in these areas to determine possible impacts.

Sensitive to change and intervention. The outcome measures selected should be sensitive to change and intervention. For example, if a rehabilitation program's participants are all severely or profoundly handicapped, then the agency should not select outcome measures such as independent living or competitive employment. Similarly, a short-term crisis intervention center should not use changes in labor market behavior or academic skills as outcome measures. In both of these examples, unrealistic measures will result in less than anticipated impacts.

Obtainable. The measures selected that meet the criterion of sensitivity to change and intervention must also be obtainable. As a program administrator, you usually have only a short time (2 to 3 years if you're lucky) to select and measure the program's outcomes. Hence, you want to look at the outcomes and impacts for which data are realistically obtainable within the alloted time. Thus, our next guideline:

Guideline 24. Don't choose a rainbow of outcome measures; choose only those that you have the time and tools to actually measure.

Table 7.1. **Exemplary Outcome Measures**[a]

Data set	Potential sources of evaluation data			
	Primary interview	Proxy interview	MIS A/E form[b]	Official records[c]
Labor market performance				
Employment	X	X		X
Job type	X	X		X
Hours	X	X		
Earnings	X	X		
Labor-force participation	X	X		
Education and training				
Attendance at school	X	X		
Attendance at training	X	X		
School curriculum	X	X		
Training curriculum	X	X		
Transfer program use				
Receipt of SSI	X	X	X	X
Receipt of OASDI	X	X	X	X
Receipt of welfare	X	X	X	X
Receipt of Medicare/Medicaid	X	X	X	X
Receipt of food stamps	X	X	X	X
Amount of SSI	X	X		X
Amount of OASDI	X	X		X
Amount of welfare	X	X		X
Living and social skills				
Living arrangement	X	X	X	
Family composition	X	X	X	
Money handling	X	X		
Transportation	X	X		
Explanatory variables				
IQ	X		X	X
Social adaptation			X	X
Academic achievement			X	X
Socioeconomic status	X	X	X	X
Prebaseline employment			X	X
Prebaseline school and training			X	X
Baseline transfer use				
Baseline living arrangement				

[a] Adapted with permission from Kerachaky *et al.* (1985).
[b] Management information system application/enrollment form.
[c] Potential sources of official records include the Internal Revenue Service, the State Unemployment Insurance program, the Social Security Administration, local referral agencies, and the Department of Labor.

Conversely, some things that are realistic may not be obtainable. The STETS demonstration, for example, focused on financial independence. It's a realistic goal, and the measures used met the first criterion. But, financial independence cannot be measured until after a person is working in a job. The STETS evaluation occurred when participants were only 7 months out of the program, which was probably too soon to determine the effects on behaviors such as doing one's own shopping or having one's own checking account. Thus, obtainability is a relevant criterion to use in deciding what to measure.

Objective measures. This criterion suggests (as do we) that you stick with objective measures that can be easily quantified and thereby measured. Some examples are shown in Table 7.1, and we give additional examples in subsequent tables. You can do objective measures on your own, and we encourage you to stick with these since they will make your life much easier. Therefore, we offer our 25th guideline:

Guideline 25.   Focus only on those objective outcomes that you can really measure.

We are aware that many programs deal with populations for which the appropriate outcome measures are difficult to objectify. For example, outcomes such as changes in quality of life, functional ability, social integration, independence, reassimilation into society, and mainstreaming into society are important and worthwhile goals, even though they do represent difficult outcomes to measure. If you focus exclusively on these, you may be beyond the scope of the book. However, many of these outcome measures are developed, and we present Table 7.2 to those readers who are producers to assist them in assessing the literature regarding their technical development and use.

Prioritized. Some outcome measures are more dominant than others. For example, if the program focuses on employment, then labor market performance will probably be the primary outcome, not independence or quality of life. Similarly, mental health programs might want to focus on symptom reduction and recidivism rather than employment or independent living arrangements. Again, one needs to use rational thought to determine which outcome measures can be used to proxy all the other potential outcomes. Table 7.1 shows how the major categories of outcome measures relate to the subsets.

III. When to Measure

We mentioned previously that impact analysis takes a long time, since it requires longitudinal data to determine the actual impact of the program upon the

Table 7.2. **Literature References to Outcome Measures for Different Program Types**

Program type	Measurement categories	Published references
Rehabilitation	Adaptive behavior level	Mayeda & Lindberg (1980); Meyers, Nihira, & Zetlin (1979)
	Client change scores	Campbell & Erlebacher (1975); Cronbach & Furby (1970); Nunnally (1975); Prosavac & Carey (1980); Washow & Parloff (1975)
	Performance indicators Work	Bellamy, Rhodes, & Albin (1985); Brown et al. (1984); Hill, Wehman, Hill, & Banks (1985); Schalock & Hill (1986)
	Community living	Baker, Brightman, & Hinchaw (1980); Landesman-Dwyer (1981); Schalock (1983)
	Quality of life	Andrews & Withey (1976); Baker & Intagliata (1982); Campbell, Converse, & Rodgers (1976); Edgerton (1975); Flanagan (1978); Schalock et al. (in press)
Education	Grades Adjustment ratings Credits Diploma	Anthony & Farkas (1982); Comfort (1982); Cooley & Bickel (1986); Havlock (1970); McDill, McDill, & Spreke (1972); Stainback & Stainback (1984)
Mental health and chemical dependency	Functional level	Carter & Newman (1976); Ciarlo, Brown, Edwards, Kiresuk, & Newman (1986); Pruchno, Boswell, Wolff, & Foletti (1983); Weisman (1975)
	Symptom reduction	Lambert, Christensen, & DeJulio (1983); Luborsky & Backrach (1974); Schneider & Struening (1983); Washow & Parloff (1975)
	Recidivism	Anthony & Farkas (1982); National Institute of Mental Health (1973); Newman & Howard (1986)

Table 7.2. (*Continued*)

Program type	Measurement categories	Published references
Corrections	Recidivism	Dolbeare (1975); Lerman (1975); Wright & Dixon (1977)
	Criminal behavior	Alexander & Parsons (1973); Davidson, Koch, Lewis, & Wresinski (1981)
Aging	Activities of daily living	Carcagno *et al.* (1986); O'Brien (1976)

participants' behavior or life-style. Therefore, one must know not only what to measure, but also when to measure it. If you have only a short time to evaluate your program, then it is best to use short-term performance indicators such as placement or admission rates. If, however, you have the luxury of time to complete multiple measures, then you must decide when to measure the outcomes. In the STETS project, for example, interview data were collected at three key points during the demonstration:

- Immediately after random assignment into the sample (the baseline interview).
- Immediately after 6 months had elapsed from an individual's random assignment into the sample, which represented a point when many experimental group members were still actively participating in the STETS services.
- At 22 months, which was well beyond the end point at which individuals stopped receiving demonstration program services.

There are no set criteria as to when outcome measures should be collected, but it is important to remember that impact analysis requires longitudinal data. Therefore, if you are a consumer reading an impact analysis, ask yourself, were repeated longitudinal measurements taken? If you are a producer, did you schedule such measurements within the constraint of the 24th guideline, that is, by choosing only those outcome measures that you had the time and tools to actually measure?

IV. How to Measure

We included data collection procedures in the three preceding chapters on participant characteristics, services, and costs. We do the same here, but first we want to stress that how one measures programmatic outcomes is tied directly to one's ability to estimate the program's impacts. For example, a short-term performance indicator can be measured to reflect the program's *outcome*. It's a fairly simple, acceptable process. But one should not imply that a short-term performance indicator reflects a programmatic *impact*. That will get you into trouble as either a producer or consumer, for as we've already mentioned, impact analysis requires longitudinal data. In the ensuing discussion on how to measure outcomes, we focus on data sources, data collection criteria, suggested data collection strategies, and the need to make some statement about the quality of the data collected.

A. Data Sources

The appropriate question to ask first is, what are all the available data sources? Refer back to Table 7.1 for a moment. This table summarizes the data elements necessary to address the STETS research issues and provides a list of potential data sources. There are five categories of variables summarized in Table 7.1. The first four (labor market performance, education and training, transfer program use, and living and social skills) reflect the areas in which STETS sought to affect the lives of program participants. The fifth category (explanatory variables) contains measures of personal characteristics and baseline measures of the outcome variables.

Data items such as those listed in Table 7.1 can be obtained from several sources, each of which has advantages and disadvantages. The primary potential sources are in-person interviews (referred to in Table 7.1 as "primary interviews"), interviews with proxy respondents (that is, persons who are knowledgeable about the primary respondent), the program's management information system application/enrollment form, and official records from several government agencies.

B. Data Collection Criteria

While each data collection strategy can provide data on the basic outcomes of interest, the quality of the data differs. Hence, here we elaborate on the six criteria presented in Table 4.1 for assessing alternative data sources and procedures:

1. *Accessibility.* A data source must be accessible; that is, the source must be willing to provide the data in a form that is usable for research.
2. *Completeness.* A crucial element in comparing data collection strategies is whether a plan can provide data for all persons in the research sample (sample completeness) and for all data items (data completeness). While

no single strategy will provide usable data on every person and every item, techniques that are seriously deficient in this respect need to be eliminated from consideration.

3. *Accuracy.* The accuracy of the research findings is related directly to the accuracy of the data. In particular, possible incentives for respondents to misreport their activities (for example, acquiescence or a desire to hide activities that are considered unacceptable) are a concern for interview strategies. For records data, the major concern lies with the data collection processes of the agencies.

4. *Timeliness.* Important features of a data source are whether it provides data that cover the period(s) of interest and how soon after such period(s) the data can be collected. While both of these issues relate to timeliness, they have very different implications for the research.

5. *Flexibility.* An important consideration in evaluating data collection strategies is the degree to which a plan can be adapted to changing circumstances. Shifts in budgets, government policies, research needs, and the quality of data sources can necessitate a change in the data collection plan. Flexible strategies provide insurance against these risks and are thus generally preferred to less flexible plans.

6. *Cost.* While cost is always a concern, the budgetary constraints of any program necessitate that it be an especially important criterion in evaluating alternative strategies.

C. Data Collection Strategies

Which data sources you use will depend largely upon the questions you are asking, the purpose of the evaluation, your program's resources, and which of the six data collection criteria are most relevant to you. The authors of the STETS project, for example, did an extensive analysis of alternative data collection strategies before actually beginning their data collection (Bloomenthal *et al.,* 1982). Based upon a thorough literature review, discussion with consultants, and pilot studies (all three of which we propose you use if you can), data for the STETS evaluation were collected from the following sources:

- Interviews with sample members and, as necessary, with proxy respondents.
- Corroborating information provided by community service agencies with which the sample members had contact and that were mentioned during the interviews.
- Background information on sample members collected by project staff as part of the intake process.
- Program participation data on all experimental group members.

- Information on program costs collected from the demonstration projects' accounting systems.
- Observations on the work activities of a sample of experimental group members.

An exercise will be helpful here. It involves completing the table shell presented in Survey 7.1. In the first column, list one or more of the potential outcome measures you would like to use (or are using) for an impact analysis or to reflect a short-term outcome. The first outcome measure might be labor market performance, with *a* being employment and *b* job type, such as you see in Table 7.1. Other prioritized headings might include functional level, symptom reduction, recidivism, criminal behavior, living standards, or activities of daily living (see Table 7.2). In the second column, indicate the potential source or sources for the evaluation data. And in the third column, evaluate each potential data source on the basis of the six data collection criteria just discussed. A simple yes or no evaluation is probably sufficient. The information you include in the survey (or don't have answers to) should be helpful to you as we move next to a discussion of the quality of the outcome measures data.

D. Quality of the Data

Whether you have already collected data, or plan to do so, you will need to know about the quality of the data. A number of techniques can be used to provide this information. A straightforward method involves simply tabulating the response rate at different data collection points, or the number of missing data sets at those same points. These techniques will tell you something about the completeness of the data, but not about its accuracy. Accuracy can be determined either by doing a validity check to insure that the data reported are accurate (as checking verbal reports of income against employer records) or by using a cross-tabs approach, wherein you compare the responses of primary with proxy respondents. An example using the cross-tabs approach is presented in Table 7.3. The basis for the numbers in each cell is the number of primary respondents who give the same answer on a particular question as do the proxy respondents.

E. Conclusion

In concluding this section on how to measure your outcomes, we would like to sensitize you to at least three pitfalls in regard to data sets and data collection. The first relates to data management. Even if all the data are perfect, there are a number of potential problems that we don't want to minimize. They include people transforming numbers, losing data sheets, misfiling data, and making errors in data entry and interpreting the printouts. Second, if you ask participants

Survey 7.1. **Table Shell of Potential Outcome Measures, Sources of Evaluation Data, and Data Gathering Criteria**

Outcome measures/ data sets[a]	Potential source of data[b]					Data gathering criteria[c]					
	Primary interview	Proxy interview	MIS form	Official records	Other	Accessible	Complete	Accurate	Timely	Flexible	Costly
1.											
a.											
b.											
c.											
2.											
a.											
b.											
c.											
3.											
a.											
b.											
c.											
4.											
a.											
b.											
c.											
5.											
a.											
b.											
c.											

[a]See Table 7.1 for an example.
[b]Place an "X" beneath the available source of data.
[c]Score each outcome measure/potential data source "yes" if the criterion is met and "no" if not met.

Table 7.3. **Cross-Tabulations for Determining Accuracy of Data**[a]

SSI receipt and amount

		Proxy respondent		
		Not received	Received, doesn't know amount	Received, knows amount
Primary respondent	Not received	60	1	5
	Received, doesn't know amount	1	0	11
	Received, knows amount	1	3	13

Welfare receipt and amount

		Proxy respondent		
		Not received	Received, doesn't know amount	Received, knows amount
Primary respondent	Not received	81	0	2
	Received, doesn't know amount	3	0	3
	Received, knows amount	2	0	4

[a]Adapted with permission from Bloomenthal *et al.* (1982).

questions, there is bias in both the questions (for example, they might be leading or ambiguous) and the answers (withheld information or acquiescence). And third, if you use data from other agencies or bureaucracies, you can encounter inconsistent definitions and sometimes inordinate retrieval delays. Thus, we want to suggest the following caution:

Caution. Don't be fooled by complete data. They may
be inaccurate due to problems in data management, response bias, inconsistent definitions, or different time
periods during which the study occurred or the data were
collected.

In addition to this caution, there are a number of points we suggest you keep
in mind in either your producer or consumer role. If a producer, so you will end
up with complete, accurate data, you should do a mini-evaluation on collecting
outcome data over a long period of time to work out the bugs that are inherent in
almost all data systems. The problem arises of course if you have a short-term
intervention program, for it is easier to get outcome measures that can potentially
be used for impact analysis from a long-term program. As suggested earlier,
short-term programs should preferably focus on short-term performance measures. As a producer, you should also be concerned about the accessibility,
timeliness, and cost criteria discussed previously. As a consumer, you should be
very concerned about data completeness and accuracy. Did the analysts, for
example, describe their data collection system? If not, worry. Did they measure
accurately what they claimed they measured? If they used objective measures, be
less concerned than if they measured more subjective outcomes such as quality of
life. Completeness is one of the stickiest issues and potential traps, since there
are two kinds of data completeness. The first concerns completeness of data
sources—whether they contain all the data you feel are important (or want, if
you are a producer). The second is *sample* completeness, which involves determining whether the data sources systematically excluded people from the data
set. Here, look for completion and differential attrition rates, as we discussed in
Chapter 4.

V. Where to Measure

In thinking about measuring outcomes, one should consider an additional
interrogatory to the three (what, when, and how) we have considered thus far.
And that is where, or in what environment, does one measure the outcome. In
this regard, we need to think about developing impact assessment procedures that
are sensitive to participant outcomes in different outcome environments. This
concept is reflected in the outcome assessment model presented in Table 7.4. As
shown in the table, one can conceptualize two types of outcome categories:
participant outcomes and outcome environments. The participant outcome and
outcome environment categories shown in the table are for expository purposes
only; different program types would substitute those outcomes and environments
that are appropriate to the respective program.

Table 7.4. **Outcome Assessment Model**

Outcome environments	Participant outcomes		
	Functional level[a]	Labor market performance[b]	Quality of life[c]
Residential	X		X
Working		X	X
Community living	X		X

[a]Examples include type of living arrangement and activities of daily living.
[b]Examples include wages, hours worked, wages per hour worked, and taxes paid.
[c]Examples for work include decreased intrusive training/supervision, integrated work, lunch and break areas, benefits, and normal work schedules. For residential and community living, examples include control of environment, social relations, and community involvement.

As an example, assume that you are the administrator of a rehabilitation program whose three programmatic goals are to increase a person's (1) behavioral skills level in community living, (2) income from work, and (3) quality of work and community life. By using the outcome assessment model outlined in Table 7.4, you can specify the types of outcomes to investigate and the environment(s) in which they are expected to occur. Subdividing the outcome environments into residential, working, and community living ensures that the expected outcomes are measured in the environment within which they would logically occur. For example, increased activities of daily living (functional level) should occur within residential and community living environments; increased labor market behavior within a working environment; and increased quality of life, in all three.

The importance of clearly defining outcome environments and corresponding participant outcomes is supported in an article by Gersten, Crowell, and Bellamy (1986) entitled ''Spillover Effects: Impact of Vocational Training on the Lives of Severely Mentally Retarded Clients.'' The study was conducted to evaluate whether the experience of working for wages would have a positive impact on individuals' lives outside the employment setting. Although no training time was devoted to social or independent living skills, the authors hypothesized that the work experience would have some impact on how the participants were perceived by caretakers and how they behaved in social settings. Across the measures used, there was no significant ''spillover'' from the working to the living outcome environments. The study indicates the importance of thinking ahead and planning carefully about both the specific outcome measures you select, and the environments in which they are measured.

Examples from two other studies further underscore the importance of the proposed outcome assessment model. The first example comes from a study

(Schalock & Lilley, 1986) that involved 85 mentally retarded persons placed into independent housing and competitive employment 8–10 years ago. The purpose of this longitudinal study was to determine the current status and participant outcomes across the three outcome environments listed in the outcome assessment model presented in Table 7.4. The participant outcomes used are identified in the top section of Table 7.5. By using a multiple outcome design, a number of potential impacts the program had on these persons could be evaluated. For example, if competitive employment is the criterion against which effectiveness is judged, then only 29% of this program's clients could be judged "successful." If independent living is the criterion, then 64% of the clients were "successful." If assessed quality of life is the criterion, then those living inde-

Table 7.5. **Example Data Sets Involving the Outcome Assessment Model**

Example 1[a]

Outcome environment	*Participant outcomes*	
Living/residential arrangement	Number who retained living placement, changed placement, or returned to the program.	
Labor market behavior	Number currently in full- or part-time employment, sheltered work, or unemployed.	
Community living	Quality-of-life indices analyzed separately for each level of living and work outcome.	

Example 2[b]

Outcome environment	*Measures characteristic of each outcome environment*	
Community placement success[c]	Family concept	Gender
	Family involvement	IPP[c] attendance
		Social-emotional behavior
	Work skills	
	Institution size	Community size
	Visual processing	
	Sensorimotor	
Program success[c]	Language	Institution size
	Sensorimotor	Visual processing
	IQ	Community size
	Education	IPP attendance

[a]From Schalock & Lilley (1986).
[b]From Schalock, Harper, & Genung (1981).
[c]See text for operational definitions.

pendently and working competitively had a higher assessed quality of life than those who were either unemployed, living with their families, or currently back in the program's residential and sheltered work environments.

The second example involved an earlier study (Schalock *et al.,* 1981) of a group of persons with mental retardation who were placed from a large institutional environment into a community-based program consisting of an array of living and work environments. The purpose of the study was to determine which measures, including institutional factors, client characteristics, intervention variables, and community characteristics, were associated with one of two outcome measures (community placement success or programmatic success). Community placement success or failure was defined on the basis of participants' either remaining in the program or returning to a state facility; programmatic success was defined on the basis of whether the client progressed to more independent housing or productive work or did not progress within either component for 2 years. These outcome environments and the specific measures associated with success within the specific environment are also summarized in Table 7.5. Thus, again we see the importance of being sensitive to measuring participant outcomes in different outcome environments.

VI. *Summary*

Measuring program outcomes is not only the first step in completing an impact analysis, but it is an essential one, since programmatic outcomes reflect what we are trying to accomplish in human service programs. We have made a distinction in the chapter between short-term performance indicators, such as placement rates, and more long-term measures, such as economic and social changes, that can be used to measure the impacts of a particular program. As we see in the next chapter, impact analysis takes a long time and involves collecting data on participants over time to see what the specific impacts of the program were upon their behavior. Thus, the outcome measures we select should ideally provide us with longitudinal change data.

We began the chapter by discussing what to measure, and we suggested that the selected measures meet five criteria. They should be attributable with reasonable certainty to the outcomes of the program, sensitive to change and intervention, obtainable, objectively defined, and prioritized. In the second section we discussed when to measure the outcomes, and we suggested that although there are no criteria for when to measure, it is important to schedule the measurements within the constraints imposed by available time and tools. In the third section we focused on how to measure the outcomes, including data sources, data collection criteria, data collection strategies, and an evaluation of the quality of the outcome data. In the last section we discussed where to measure the outcomes, and we presented an outcome assessment model that relates participant outcomes to different outcome environments.

We recognized throughout the chapter that most readers are consumers or producers of small outcome evaluations. In that regard, we presented a number of guidelines:

- Outcome measures should be tied to a program's goals and objectives; keep in mind the comparisons you are making.
- Don't choose a rainbow of outcome measures; choose only those that you have the time and tools to actually measure.
- Focus on only those objective outcomes that you can really measure.

Additionally, we stressed that measuring program outcomes and impact analysis is both difficult and complex and involves sophisticated techniques and procedures that are beyond the scope of this book. Thus we have taken a "broad brush" approach and suggested some techniques to use if you are a producer, but we also cautioned that the suggested procedures will result in estimates with substantial uncertainty. Therefore, we offered the following cautions:

- The techniques we present result in only a partial impact analysis. Impact analysis involves certainty and precision; what you will have from our techniques is a rough idea.
- Don't be fooled by complete data. They may be inaccurate due to problems in data management, response bias, inconsistent definitions, or different time periods during which the study occurred or the data were collected.

In Chapter 8, we pursue further the difference between outcomes and impacts. Outcomes are what happens as a result of some program activity, whereas impacts refer to how the outcomes differ from what would have happened if the program was not there.

VII. Additional Readings

Anothy, W. A., & Farkas, M. (1982). A client outcome planning model for assessing psychiatric rehabilitation interventions. *Schizophrenia Bulletin, 8*(1), 13–38.

Fairweather, G. W., & Davidson, W. S. (1986). *An introduction to community experimentation: Theory, methods and practice.* New York: McGraw-Hill.

Hargreaves, W. A., & Attkisson, C. C. (1978). Evaluating program outcomes. In C. C. Attkisson, W. A. Hargreaves, M. J. Horowitz, & J. E. Sorensen (Eds.), *Evaluation of human service programs* (pp. 303–339). New York: Academic Press.

Kish, L. (1967). *Survey sampling: Methodology survey and evaluation.* New York: Wiley.

Lansing, J. B., & Morgan, J. N. (1971). *Economic survey methods.* Ann Arbor, MI: Institute for Social Research, The University of Michigan.

Mahoney, M. J. (1978). Experimental methods and outcome evaluation. *Journal of Consulting and Clinical Psychology, 46*(4), 600–672.

Newman, F. L., & Howard, K. I. (1986). Therapeutic effort, treatment outcome and national health policy. *American Psychologist, 41*(2), 181–187.

Shonfield, A., & Shaw, S. (Eds.) (1972). *Social indicators and social policy.* London: Heinemann.

Washow, I. E., & Parloff, M. B. (Eds.) (1975). *Psychotherapy change measures.* Washington, DC: U.S. Department of Health, Education, and Welfare. National Institute of Mental Health (Pub. #ADM-74-120).

8

Estimating Program Impacts

I. Overview

We talked about comparisons in the previous chapter and indicated that evaluation involves structured comparisons. Now the issue is really before us. This is the point where we get to the tough issues we mentioned regarding the need to measure outcomes on both sides of the comparison (participants and nonparticipants) and then contrast these outcomes. In the simplest and most ideal sense, impacts of a program are estimated by contrasting the outcome measures for a group of participants with those for a comparison group. Administrators need to make many decisions that involve comparisons and choices among alternatives. Throughout the book we have stressed that administrators need to make these decisions based on valid evaluation data, and that evaluation involves structured comparisons. In this chapter, we discuss various ways to structure comparisons and thereby be in a better position to estimate program impacts.

We are going to structure our discussion around making a specific type of comparison: between a new program and the status quo, or what would happen if the program had not been introduced. We feel that this type of comparison is the most policy relevant, since we've observed that people frequently want to provide new, coherent, focused services to replace the mix of services already available. Hence, the policy choice and decision is frequently between a new program and the status quo. A good example is the current emphasis on using supported employment programs for persons with developmental disabilities to replace the available mix of sheltered workshops and day activity centers.

The challenge facing the reader as either an evaluation producer or consumer is to measure in a consistent way values of the outcomes we discussed in Chapter 7 under each situation or condition. This process is diagrammed in Figure 8.1. The figure shows the ideal situation in which equivalent groups are formed: one group receives the intervention, and one does not. If outcomes are measured for both groups and differences found, then for both the producer and consumer, the dilemma faced is whether the mean differences are due to the intervention. The ideal situation depicted in Figure 8.1 is further compounded by

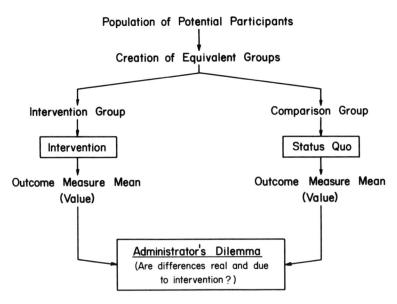

Figure 8.1. Steps involved in determining a program's impacts.

the fact that although many programs have consistent measures from program participants, they frequently do not have measures reflecting what the participants would have done had the program not been there. And furthermore, there is a strong probability that when all is said and done, one will not find significant impact effects due to either random causes or ineffective interventions; that is, for some reason, the program didn't work. Hence, in planning an impact analysis, one needs to think ahead and consider a fallback position, such as barrier analysis. If one knows which barriers are preventing significant impacts, then in the next replication, one can plan ahead to overcome them.

Our intent in this chapter is to focus primarily on the consumer who is attempting to evaluate someone else's impact analysis. However, we will also examine in more detail two techniques that can be used by an administrator with a limited budget. We encourage those who can conduct a limited impact analysis to do so, but within the cautions and guidelines we provide. We feel it will be a reasonable return on investment. However, we repeat our previous caution that we are using a broad brush approach to outline some specific, but rough, impact analysis procedures that result in impact estimates with substantial uncertainty. We further caution the reader that the procedures we outline result in only a partial impact analysis due to probable uncertainty and imprecision. But again, we don't want to scare off the reader. Our suggested techniques should result in better statements about your program's potential in meeting those goals and objectives you described in Chapter 2 regarding your setup and program rationale.

The chapter is divided into four sections: a discussion of four procedures that can be used to estimate the effects of a program; a proposed impact analysis model; some practical impact analysis techniques; and a brief discussion of three validity issues surrounding impact analysis. As in the previous chapter, we use examples from the STETS demonstration to illustrate important points regarding impact analysis. Our concern is that the reader may get lost in some of the details we discuss, so our first guideline reiterates the general procedure for estimating program impacts.

Guideline 26. The general procedure for estimating the impacts of a program is to compare the outcome measures for those who receive the intervention with those who do not.

II. Estimating the Effects of the Program

An impact analysis deals with comparisons of how program outcomes differ from what would be the case in the absence of that program. Many comparisons can be made, and we suggest the most policy-relevant one is between a new program and the status quo. Measuring the program's outcomes is relatively easy, as we saw in Chapter 7. The challenge is to get consistent measures of the status quo. In this section, we review four evaluation designs that can be viewed as ways of getting consistent, accurate measures of what would happen under the comparison situation.

To estimate the effects of a human service program, an analyst must determine what the participants actually did and what they would have done in the comparison situation. While there are numerous measurement issues, the first step in this process is relatively straightforward. It requires only that the analyst observe what the participants actually did. The second step is conceptually more difficult, since the analyst must determine what the participants would have done under different conditions. This comparison situation cannot be observed; it must be estimated on the basis of factors that can be observed.

In this section we discuss four ways to estimate what a group of participants would have done under alternative conditions. In general, this involves studying the activities of a comparison group—that is, a similar group that was not offered the program services being evaluated. This similar group can be identified using the methods of classical experimentation or through the use of various matching techniques. Regardless of which method is used, the essential goal is to find a group that is identical to the group enrolled in the program, with the exception of receiving special services. When this goal is achieved, the differences that emerge between the two groups can reasonably be attributed to the effects of the program under study.

A. Experimental Design

The methods of classical experiments generally provide the most accurate and statistically valid means of identifying a comparison group. These methods randomly assign program applicants to two groups (see Figure 8.1). One group is offered the services that are being evaluated; the other group is offered only the services available under the comparison situation. If the assignment to these two groups is random, the two groups should be identical in all respects except for the receipt of the services being studied. However, as with all random events, a chance exists that the two groups will differ in some respect. The probability of this chance diminishes as the sample of persons who are randomized becomes larger. On average, the two groups should initially exhibit the same measurable characteristics (for example, age, gender, ethnicity, previous schooling, and residential situation) as well as the same unmeasurable characteristics (for example, motivation, work attitudes, and general opportunities).

The advantage of an experimental design is that if the number of persons assigned is moderately large, the analyst can be reasonably sure of the comparability of the two groups. In particular, the comparability of the groups in terms of unmeasurable characteristics is important, since it is very difficult to control for the influence of such characteristics using statistical methods. Furthermore, experience has shown that results based on data generated from an experimental design tend to be stable with respect to changes in the specific details of the estimation process. Thus, this design can produce results in which a relatively high degree of confidence can be placed.

Although the experimental design is used in physical science and medicine, it hasn't caught on very well in human services for reasons discussed in the next paragraph. It is obviously more appropriate for programs that are in either the demonstration or ongoing stage of development rather than one in the feasibility phase. If, however, you have 50 openings and 100 applicants and a year's time, then you can randomize as Figure 8.1 reflects. But don't attempt this procedure if it involves denying services, withholding treatment, or using a small sample. A small sample will wipe out any effects, and thus the obtained mean differences could be a fluke.

While experimental designs have been used for some social programs, they are not always feasible. For example, it may not be possible to conduct random assignment in an entitlement program. Such programs guarantee services to all members of a specific target group, and so it would be infeasible to assign some applicants to a control group. However, if all applicants cannot be served or if the intervention is a new one in which there is still doubt about its effectiveness, then random assignment can be a fair way of deciding who should get the program services, in addition to being a great aid in the analysis.

B. Comparison Group

When it is infeasible to use an experimental design, there are other means of identifying a comparison group. Such a comparison group should resemble the group under study and conform with the specified comparison situation. For example, if the comparison situation is a specific alternative program, the analyst would look for a group of persons who were being served by that alternative program. Such a group might occur within the same or a different community. Other comparison groups that have been used by researchers include persons who were offered program services but who did not actually enroll and persons in similar circumstances in other geographical areas where the special service is unavailable.

The problem with the comparison group approach is that the treatment group (the persons who get the special services) and the comparison group may differ in ways that confound the analyst's ability to isolate the effects of the intervention. For example, differences in individual characteristics, economic and social opportunities, parental support, ability, or motivation, if correlated with participation in the program under study, may create differences between the treatment and comparison groups that have nothing to do with program impacts. Analysts often use statistical methods, such as regression analysis, to control for these differences between the comparison and treatment groups. However, this approach is problematic because the analyst never really knows whether all factors that potentially influence the outcomes have been accounted for appropriately and, therefore, whether the estimated impacts of the program are reliable. Such problems arise because of difficulties in (1) identifying precise or accurate measures of the specific characteristics that the analyst wants to control for, (2) including all relevant factors in the statistical model, and (3) understanding and specifying the appropriate relationship among the factors and the outcomes of interest. All of these problems, if not handled appropriately, can create biases in the statistical estimates of impacts.

C. Hypothetical Comparison Group

A third approach to identifying a comparison group is to use conjecture. By relying on a general knowledge about the average outcomes of nonparticipants or on a knowledge of preenrollment status, the analyst may estimate what would have happened to participants had they not enrolled in the program. Some researchers of supported work programs (Hill *et al.,* 1985), for example, have estimated impacts under the assumption that had participants not enrolled in the program they would have continued in the activities they had prior to enrollment. These methods clearly represent very crude estimation procedures and, ultimately, are not likely to withstand rigorous analytical criticism. These methods also require the use of sensitivity tests to reflect what would happen if other assumptions were made.

This third approach, however, can be useful in assessing the potential of programs to generate benefits that will outweigh costs, particularly when the likely pattern of participant activities in the absence of the intervention is well known. For example, the approach might work well for evaluating programs that teach profoundly disabled children. In this case, history indicates that such children have faced a very narrow range of employment, social, and residential options. Thus, an analyst might have a reasonable degree of success in specifying an appropriate comparison situation for a program that serves such students.

However, this approach is likely to be misleading for groups that have a wider range of options. Kerachsky *et al.* (1985) investigated this issue when they examined the activities of a randomly assigned control group in the STETS demonstration. They found that the percentage of control group members who held regular jobs rose over the observation period, so that 20% of them held such jobs at the end of the 22-month study. Thus, had the study been limited to using the difference between the preprogram and postprogram behavior of the participant group to estimate the impacts, it would have estimated substantially larger net program impacts than actually occurred.

D. Pre–Post Measures on the Same Group

This is a popular method in which persons are assessed on one or more potential outcome variables before and after the intervention. Any changes are then attributed to the intervention. A typical example is that reported by Aninger and Bolinsky (1977), who evaluated the changes in adaptive behavior in 18 retarded adults transferred from an institutional setting to supervised apartment living environments. Pretest measures were made before placement and 6 months after placement. This comparison technique poses a number of threats to the study's internal validity, as discussed later in the chapter.

E. Conclusion

The choice among these four approaches for estimating impacts—experimental design with randomized treatment and control groups, comparison group, conjecture, and pre–post measures on the same group—will depend on the degree of certainty needed by policymakers. In general, the experimental design will produce the most accurate estimates, and it should be used whenever feasible. In those cases in which it is impractical to use an experimental design, the comparison group method often provides a reasonable alternative, if it includes rigorous statistical controls for intergroup differences that might influence the outcomes of interest. The conjectural and pre–post approaches should be used only when the primary interest is to develop hypotheses about the potential performance of a program; estimates based on these approaches are unlikely to provide an accurate test of a program's impacts.

The relationship among these four approaches to estimating impacts, the assumptions made, and the study's precision are diagrammed in Figure 8.2. As shown in the figure, certainty and precision (internal validity) increase as one moves from pre–post to experimental techniques. Conversely, the assumptions that one makes decrease for the experimental and comparison group techniques and increase for the hypothetical comparison and pre–post techniques.

In summary, the certainty and precision regarding program impacts depend greatly on the comparison technique used. This reflects a guideline that pertains to producers and consumers alike:

Guideline 27. The certainty and precision of impact statements depend upon the comparison technique used. An experimental design results in the greatest certainty and precision.

Certainty	Precision (Internal Validity)	Comparison Technique	Assumptions
HIGH	HIGH		LOW
		Experimental Design	Carefully selected comparison group; few assumptions made.
		Comparison Group	
		Hypothetical Comparison Group	Assumptions made, but sensitivity tests done to see the effects of different assumptions.
		Pre-Post on Same Group	
LOW	LOW		HIGH

Figure 8.2. Relationships among comparison techniques, assumptions made, and study's precision.

III. Impact Analysis Model

Before presenting our impact analysis model, we would like to sensitize you to some of the actual data and corresponding discussion from the STETS impact analysis. The estimated program impacts on key outcome measures are summarized in Table 8.1. Note the following when you read the table. The outcome measures are those we discussed previously in reference to Table 7.1. The three time periods (6, 15, and 22 months) represent those postenrollment times when data were collected (using the data collection procedures discussed in Chapter 7). The experimental group represented the program participants, and the control group, those who were randomly selected for nonparticipation status. And note, finally, the essence of impact analysis:

Experimental group mean − Control group mean = Estimated impact

As can be seen, some of the differences between the two means are significant (represented by an asterisk), and some are not. Employment in regular jobs was significantly greater for experimental group members than for control group members, and by Month 22, experimentals were an average of 62% more likely than controls to be employed in a regular job. A significant increase in average weekly earnings was also seen in the experimental group, as was a significant decrease in the percentage of experimental group members in training. An interesting finding was that the significant impacts related to schooling and some transfer uses disappeared by the end of the demonstration period, which reinforces our earlier statement regarding the need in impact analysis for longitudinal data.

Because of the substantial impact on the incidence of regular job holdings, hours worked, and earnings, it might be expected that the STETS demonstration would also have impacts on other areas of participants' lives—especially their overall economic status, their independence in financial management and living arrangement, their use of formal and informal services, and their general level of involvement in regular, productive activities. Rather than present detailed tabled data, we simply summarize the findings.

The expected direction and duration of the effects of the program on economic status, measures of independence, and life-style were not always clear. Several factors in particular cloud the results of the evaluation of these impacts. First, the increased earnings observed for experimentals appear to be partially offset by decreases in transfer benefits and other sources of income, thereby diluting the overall financial impacts of the program. Second, although STETS may have had impacts on financial management skills and independent living arrangements, those impacts may follow others with a considerable time delay, in which case the 22-month observation period was too short to observe them.

Table 8.1. **Estimated Program Impacts on Key Outcome Measures**[a]

Outcome measures	Month 6			Month 15			Month 22		
	Experimental group mean	Control group mean	Estimated impact	Experimental group mean	Control group mean	Estimated impact	Experimental group mean	Control group mean	Estimated impact
Employment									
Employed in regular job (%)[b]	11.8	10.7	1.1	26.2	16.8	9.4**	31.0	19.1	11.9**
Employed in any paid job (%)	67.8	45.2	22.6*	44.8	43.6	1.2	44.7	43.7	1.0
Average weekly earnings in regular job	$11.81	$9.81	$2.00	$26.90	$16.31	$10.59**	$36.36	$20.55	$15.81**
Average weekly earnings in any paid job	$52.39	$25.93	$26.46**	$37.91	$26.48	$11.43**	$40.79	$28.41	$12.38**
Training and schooling									
In any training (%)	61.7	40.6	21.1**	20.60	28.4	-7.8*	16.6	29.1	-12.5**
In any schooling (%)	7.5	15.7	-8.2**	6.2	10.1	-3.9	8.0	11.4	-3.4
Income sources									
Receiving SSI or SSDI (%)	26.3	31.0	-4.7	33.1	40.7	-7.6**	34.9	40.2	-5.3
Average monthly income from SSI or SSDI	$66.41	$74.59	-$8.18	$91.35	$109.65	-$18.30	$99.27	$120.03	-$20.76
Receiving any cash transfers (%)	31.7	43.1	-11.4*	44.5	51.5	-7.0*	49.6	52.0	-2.4
Average monthly income from cash transfers	$80.23	$99.98	-$19.75	$114.78	$138.72	-23.94	$126.53	$136.08	-$9.55
Average weekly personal income[c]	$71.72	$50.94	$20.78**	$67.22	$59.67	$7.55	$71.59	$62.39	$9.20

Note. These results were estimated through ordinary least squares techniques.

[a] Adapted with permission from Kerachsky *et al.* (1985).

[b] Regular jobs are those that are neither training/work–study nor workshop/activity center jobs.

[c] Personal income includes earnings, cash transfer benefits (AFDC, general assistance, Supplemental Security Income, and Social Security Disability Insurance), and other regular sources of income.

*Statistically significant at the 10% level, two-tailed test.

**Statistically significant at the 5% level, two-tailed test.

Third, although the program generated increased earnings for sample members, those increases might not have been enough to enable them to live independent life-styles—especially in such large metropolitan areas as New York or Los Angeles. Finally, parents and counselors might have wished to see more concrete and stable earnings gains before they were willing to give the sample members greater independence.

Despite these limitations in the ability to detect what may be primarily long-run effects, the study pursued an analysis of such impacts due to the strong policy interest in understanding the effects of transitional employment programs on the life-styles of participants. In general, they observed some relatively small program effects on such measures of independence as overall economic status, services received from community agencies, and involvement in activities oriented toward employment. However, these effects generally declined to a great extent in the postprogram period, seemingly due to two factors. First, in the later observation periods, either the direct effects of STETS participation on such outcomes as total income, service utilization, and level of inactivity were no longer evident, or, where they were evident (as with personal income), the estimated effects were not statistically significant. Second, while the STETS experience provided a head start toward independence for many sample members, those who did not participate in the program (controls) also began to achieve similar levels soon afterwards. Although certain subgroups (for example, those with a moderate level of retardation) did seem to continue to benefit from the program, even those who were more likely to achieve and maintain positive effects from their experience in STETS exhibited relatively low levels of independence by the end of the observation period. Most of the mentally retarded young adults in the study had total personal incomes that were less than the poverty level, thereby restricting their opportunities for achieving a more economically self-sufficient life-style. Possibly because of their low incomes, most continued to live with their parents and to exercise little independence in financial management. Finally, a substantial proportion of both the experimental and control groups were not involved in employment, school, or training, and thus had limited opportunities to gain valuable work experience and skills.

Thus, given the short postprogram period for which data were available, one cannot tell whether a more economically and socially independent life-style would eventually be achieved by the participants, or whether the effects of participating in a transitional employment program would become more evident at a later period. This statement often reflects the reality frequently found in even a large, well-funded impact analysis: despite rigorous experimental methodology, there can still be uncertainty in the impact measures obtained.

Now that you have a feel for impact analysis, we would like to outline an impact analysis model that should assist your producer or consumer efforts. The

$$(T_F - C_F) \quad - \quad (T_B - C_B)$$

Differences After   Differences Before

Figure 8.3. Impact analysis model: Differences attributed to the program.

model we propose is diagrammed in Figure 8.3. The abbreviations used in the model refer to treatment or intervention T, comparison C, following intervention F, and baseline or before intervention B.

For expository purposes, assume that an administrator of a program dealing with highly aggressive persons wants to determine the impact of a particular drug on the aggressive behavior of persons committed to the program. Behaviors selected for analysis include physical assaults on others, as reflected in hitting, biting, kicking, and scratching. An assault index is used to report the number of these behaviors per day. Before the intervention begins, the index is completed for each person for the pervious 3 months. After the baseline information is collected, each person for whom consent is obtained is placed into either a drug intervention (T) or control/comparison (C) group. The assault index is tabulated daily for each individual during the 3-month study. In reference to the model, one computes the differences between the groups before intervention, ($T_B - C_B$), and subtracts it from the differences after, ($T_F - C_F$). Once the differences are obtained, they are tested for statistical significance, from which impact statements can be made.

The proposed impact analysis model can be used to conceptualize what might influence a program's effects. For example, if you look only at postintervention effects, ($T_F - C_F$), it is easy to overlook initial differences and the fact that groups could differ at baseline. It may be, for example, that some individuals in the placebo (or drug) group had higher baseline assault rates than others. The model sensitizes you to those baseline differences. If, on the other hand, you focus only on pre–post comparisons, ($T_F - T_B$ or $C_F - C_B$), it is easy to overlook an important threat to internal validity; the participants in either group or both might have changed over time due to their maturation or history. A related concern is whether the groups were initially the same. That concern is handled easily by simply evaluating the baseline difference ($T_B - C_B$).

Let's now relate the model to the four comparison techniques discussed in the previous section. As a prelude, remember that we want to compare change under the treatment condition with that under the comparison condition. Thus, we use one of the comparison techniques to determine the comparison group difference. But first, a caution:

> *Caution:* If you use the impact analysis model, don't forget that other factors are occurring that are not reflected in the simple model.

Table 8.2 presents the format for our discussion of the relationship between the impact analysis model and the four previously discussed comparison techniques (see Figure 8.2). The first column lists the comparison method; the second, what part of the equation to analyze to determine the significance of the program's effect(s); and the third column summarizes the assumptions one is making regarding conditioning factors (participant characteristics that affect the outcome) and baseline levels on the outcome variables. If one uses classical experimental method, at baseline the two groups are identical, $T_B = C_B$, and the entire expression (and the potential program impacts) can be reduced to $T_F - C_F$. With the comparison group method, you need to analyze differences reflected in the model: $(T_F - C_F) - (T_B - C_B)$. Additionally, you have assumed the effects of any unmeasured differences between groups are small. With the hypothetical comparison technique, one analyzes $T_F - T_B$ and assumes that $C_F - C_B = X$, where X represents conjecture that the comparison group would have changed along the same line as the treatment group. And finally, with the pre–post method, you again analyze $T_F - T_B$ and assume that $C_F - C_B = 0$.

Table 8.2. **Use of the Impact Analysis Model to Demonstrate the Relationships among Comparison Techniques, Program Effects, and Assumptions**

Method	Program effect	Assumptions		Certainty
		Conditioning factor	Baseline outcome variable	
Experimental	$(T_F - C_F)$	Equal across groups (by design)	$T_B - C_B = 0$ (by design)	High
Comparison group	$(T_F - C_F) -$ $(T_B - C_B)$	Assume equal across groups (or control statistically)	Controlled for $T_B - C_B$	
Hypothetical	$T_F - T_B$	Assume equal	Assume $C_F - C_B = X$ (X = conjecture that comparison group would have changed along same line)	
Pre–post	$T_F - T_B$	Assume equal	$C_F - C_B = 0$	Low

Our advice is to not get caught up in all the "how to's" of the model, for we have barely scratched the surface. We include it to demonstrate the nature of the assumptions you have to make as you fall back from the randomization/ experimental model. For example, an enormous range of activity takes place in what we referred to as a comparison group. The consumer needs to know when a comparison group is used and understand why you are trying to come as close as possible to what would have occurred if the true experimental/random design had been used. With a comparison group, you are trying to use a variety of matching and statistical techniques to control for, eliminate, or filter out all differences between groups except those due to the treatment. As a consumer who is reading someone else's analysis that included a comparison group, you need to be convinced that the residual is due to the program. With randomization, you assume it is.

As producers, be cautious about making claims about programmatic impacts that are based on methods containing considerable uncertainty and assumptions. Small programs are extremely unlikely to have the time or resources to use large-scale randomization or comparison methods. You are better off to use a pre–post method and see where that takes you. If it looks promising, then you might be willing to commit the resources necessary for a large-scale effort. Or, as we see in the next section, you might be able to use the hypothetical comparison method, in which you can use a number of alternative assumptions regarding the comparison group. The analogy we propose is that of a hiker. If you are hiking with a map and come upon a swamp or rough mountains, you have two options: if you are an expert, it's probably safe to traverse either; but if you are a novice, it's better to walk around the danger, even though it might take you a bit longer.

In summary, we hope the impact analysis model helps you to conceptualize and understand what is meant by a program's effect, or impact, and how assumptions affect certainty and precision. The basis of the model is that one observes differences in two groups and wants to attribute those differences to the program or intervention. Thus, one observes $T_F - C_F$ after the intervention, and wants to know how much of the difference is due to the program. With random assignment, used in the experimental method, probably all the difference can be attributed to the program. With the comparison method, you need to control for initial differences between the two groups; but unfortunately, the only things you can control for are what is measured, and you need to assume that what was measured captures all the differences. With the hypothetical comparison, you assume that some defined outcome would have happened, and with the pre–post techniques, you assume that in the absence of the intervention, nothing would have happened. Therefore, be aware that in using the impact analysis model, you are making assumptions about the groups. Thus, another caution:

Caution: It is okay to use the impact analysis model for exposition and as a guide; but remember that it is weakened due to factors that are unmeasured and the assumptions made.

IV. Practical Impact Analysis Techniques

Much of this chapter has focused on the evaluation consumer, not the producer. We have done so purposefully since this book is a field guide, and once you get into the comparison techniques summarized in Figure 8.2, design and data collection issues will cause you significant problems. In fact, if you venture into that morass, there may be more problems than payoffs. But we did promise to provide for producers two techniques that are appropriate for feasibility-level programs. They include using either the hypothetical comparison method or performance indicators.

A. Hypothetical Comparison

This can be a valuable technique that results in the ability to make some impact statements, but it can be used only if you have follow-up data on the participant group T_F. In using this technique, you are estimating what would have happened to participants had they not enrolled in the program, by relying on a general knowledge about the average outcomes of nonparticipants or on a knowledge of preenrollment status. The example we use is that of Hill *et al.* (1985), who estimated impacts under the assumption that had participants not enrolled in the program, they would have continued in the activities they had prior to enrollment. This method becomes more appropriate if the choices of activities of a group are constrained and one knows from historical evidence what they will do.

The analysis that the authors report assessed the impact of the supported work model in placing persons with mental retardation into comptetitive employment. This was a time (1978–1984) when supported employment was a new idea and its merits doubted by some. The investigation did not include a comparison group, but since the analysts had considerable data on participants, they wanted to know the effect of the supported work model. Thus, they developed the hypothetical, conjectural comparison group. Part of their rationale was that most of the participants had been considered by service providers as not ready for competitive employment, had never worked before, and were referred only on the basis of needing intensive, long-term job site training. Thereby, their confidence was increased that few participants would have been sustained in competitive employment without supported work services, and they *assumed* that the participants would continue to do the same thing they did before they entered the supported employment program.

The three primary outcome measures the authors used were increased monetary output, decreased use of alternative programs, and decreased government subsidies (see Table 7.1 for comparable measures). Impacts were based on longitudinal data collected on 155 persons with mental retardation who were placed into part- and full-time competitive jobs from the fall of 1978 to 1984. The participants' age range was 18–63, with an average measured intelligence score of 50 (range, 27–78). At the time of placement, 86% were receiving Supplemental Security Income (SSI) or other government subsidy. Only 18% had earned over $200 in annual salaries from sheltered or nonsheltered employment the year prior to placement.

Data used for the impact estimates came from two sources: actual participant income and secondary data sources. For example, fringe benefits were estimated based on data from the U.S. Department of Labor; participant taxes paid were estimated by using an effective tax rate (23%) applied to the estimate of increased net income; decreased use of alternative programs was estimated based on a previous state study of the average cost of day adult service programs; and decreased government subsidy was derived by computing actual SSI reductions due to each participant's earned income over the period of employment. The analysis showed that the supported employment model had potential, and even though a very simple study was done, it had a significant impact on policy.

Note that numerous secondary data sources were used by the analysts in computing the estimated impacts. We identify many of these sources in Chapter 9. For the time being, it is important to realize that one doesn't need all the necessary data in the program's management information system to do this type of impact analysis. But it is essential to realize that the results are based upon the assumptions made and the secondary data used (such as taxes paid and alternative program costs). What would be the result of using different percentages and costs, for example? Thus, we propose a very important guideline regarding the use of sensitivity tests, if you use hypothetical comparison groups (we discuss sensitivity tests in detail in Chapter 9).

Guideline 28. When you make assumptions in your impact analysis, and there is uncertainty in regard to those assumptions, sensitivity tests should be conducted to get a handle on the implications of the assumption(s).

B. Performance Indicators

Sometimes, program administrators have no follow-up data. What can they do then? Our suggestion is to try to use one or more performance indicators, such as placement rates, graduation rates, successful terminations from the program, or admission rates to other programs or facilities, to estimate the program's effects. Begin with the setup and rationale, and look for outcomes that are short term and closely related to the outcomes in which you are interested. The good news is that performance indicators are frequently literature based and can thereby be used for comparisons (we summarize many of them for you in Chapter 9). For example, if your program is attempting to affect employment placement rates, use those rates and compare them to rates found in the literature from similar programs to get some idea about exposure to employment. The draw back to using performance indicators is that you may really want to determine the impact of your program on long-term employment; and if you use placement rates, be aware that you have not looked at long-term effects. Thus, a caution:

Caution: Remember that performance indicators, such as placement rates, are a necessary but insufficient condition in impact analysis.

V. Validity Issues in Impact Analysis

We realize that this is not a textbook on statistics; but we also feel that as either a producer or consumer, it is important to understand three validity issues in impact analysis. In reference to impact analysis, to be *valid* means to be logically correct in one's estimation of impacts. The three validity issues that affect that logical correctness include statistical, internal, and external (generalizability) validity.

A. Statistical Conclusion Validity

The purpose of conducting a statistical test is to determine whether the differences in the outcome measures between groups are likely to be due to chance or nonchance factors. If one says that the differences are significant, then one infers that the treatment or intervention made the difference. All things considered, the larger the intervention effect (as frequently reflected in the difference between group means), the more confident one can be that the intervention or treatment had an effect on the outcome measure. However, one must also take into consideration the amount of variation in the outcome measure. Most statistical tests analyze the ratio of the variance (or differences) between the groups to variance (or differences) within each group. Statistical significance is reached when the between-group differences exceed the within-group differences to an extent that they cannot be considered due to chance fluctuations. By convention, most investigators use either the 5% or 1% significance level, which is depicted as $p < .05$ or $p < .01$. This depiction means that the null hypothesis (no difference) is rejected and the alternative hypothesis that the intervention produced the effect on the outcome is accepted.

In reference to impact analysis, calculating experimental–control group differences simply in terms of the mean values of outcome variables may produce biased estimates of intervention or treatment effect, especially if there are differences among preassignment characteristics. Hence, regression techniques are frequently used. These techniques are advantageous because they control for sample differences of this type, and they can be expected to produce unbiased estimates of intervention effects. They also offer two other advantages over a simple comparison of mean values. First, regression analysis provides more powerful tests of the program's potential effects because it allows one to control statistically for the influence of other explanatory variables. Second, by including the explanatory variables in the regression model, one can directly assess their individual net influences on the outcome variables within a simple analytical framework.

B. Internal Validity

The impact analysis model allows one to conceptualize the outcomes from—and potential changes produced by—a particular program. The concept of statistical conclusion validity results in statements regarding the statistical significance of differences in those outcomes. At this point, another impact analysis issue arises. The issue involves estimating the impacts to similar groups and settings. For example, in reference to the exemplary drug intervention study, if significant differences were obtained and the drug reduced assaultiveness, could one estimate similar effects in comparable groups. The issue of estimation is an important one for administrators and policymakers alike. Administrators want to demonstrate that their program produced the significant impacts, which is the issue of internal validity. On the other hand, policymakers need to know which programs are effective so that successful program models, techniques, or intervention strategies can be replicated elsewhere, which is the issue of generalizability, or external validity.

Internal validity refers to the authenticity of treatment or intervention effects. It deals with questions such as, did what you do really produce the effect? or, can you really support your conclusions about the program's impacts? The degree of confidence in supporting those conclusions increases if one randomly assigns subjects to groups so that groups are equated initially; then significant posttreatment effects can be attributed to the intervention received. But one's confidence in supporting the conclusions can also be increased by being sensitive to, and controlling, a number of threats to internal validity that weaken the authenticity of the reported effects. These threats, summarized in Table 8.3, are history, attrition, maturation, regression toward the mean, and measurement effects.

a. History.   This involves events that occur at the same time as the intervention. Ideally, the treatment and comparison groups have the same history during the evaluation period, with the only difference being the treatment or status quo conditions. But that is seldom the case. This threat to internal validity occurs frequently in placement studies in which subjects make new friends, interact with a larger group, and are indirectly subjected to vocational, recreational, and social "interventions" that accompany placement. Hence, a differen-

Table 8.3. **Threats to Internal Validity**[a]

Threat	Description
History	
Events extrinsic to research design	Events occur coincidentally with intervention
Events intrinsic to research design	Procedural variations
Attrition	Loss of subjects due to mortality, upward mobility, or downward mobility
Maturation	Growth, degeneration, warmup, and fatigue that impact on the dependent variable(s) regardless of any intervention
Regression toward the mean	The statistical tendency for individuals with extreme scores on one occasion to achieve more nearly average scores on subsequent occasions
Measurements	
Reaction to assessment	Measurement of the dependent variable(s) creates the illusion of a treatment effect where none exists in reality
Biased testers or observers	Examiners whose scores are based on information other than that provided by the subject
Floor or ceiling effects	The range of a measure is constrained so that the performance of a high-scoring subject is underestimated or that of a low-scoring subject is overestimated.

[a]Adapted with permission from Heal and Fujiura (1984).

tial history of the two groups can have a direct effect on the observed outcomes. Similarly, variations in the procedures used over time will significantly affect internal validity. A common tendency in human service programs is to modify the intervention as one gains information about what works and what does not. Although this is laudable, it does pose a threat to internal validity, for it contaminates the purity of the treatment or intervention.

b. Attrition. Differential dropout rates can result in measures being taken on different populations. This threat, which we discussed in considerable detail in Chapter 4, is common in job-training, drug rehabilitation, remedial education, and correction programs.

c. Maturation. Human growth and development is a continuous process. It is reasonable to assume that many adaptive and social skills develop even in adults that are independent of a potential intervention. Thus, programs dealing with rehabilitation and education need to be especially sensitive to this threat to internal validity, since participants may change due primarily to maturation as opposed to the assumed intervention.

d. Regression toward the Mean. Regression toward the mean is a statistical concept that refers to the drift of scores on an unreliable measure. Its effect is to cause individuals who have extreme scores on the first testing to have scores that are less extreme on subsequent testings. This threat is also common in placement or parole studies in which persons selected frequently are those with the highest adaptive or social skills. Therefore, one would expect some regression in their scores after they have been placed, even if their real skill level has been stable.

e. Measurement. A number of potential threats to internal validity can result from characteristics of the measurement device or technique. For example, if subjects are interviewed, they can practice or give the responses they perceive the interviewer desires. Another threat pertains to a bias that comes from the failure to use blind observers, double-blind procedures, or to assess interobserver agreements. A final measurement problem deals with a floor or ceiling effect in which the range of a measure is limited (because of a person's extremely high or low pretest score) and therefore the performance of a high-scoring subject is underestimated and that of a low-scoring subject overestimated. Measurement error problems, and their threats to internal validity, occur across human service programs and studies dealing with their effects.

At this point, we realize that you might be overwhelmed with all these threats to internal validity, and might well be asking yourself, Is it either worthwhile or possible to do impact analysis? Our answer is yes to both; but we caution you to be aware of these threats that tend to decrease the certainty and precision of your impact statements.

C. Generalizability

Administrators and policymakers are frequently concerned with the extent to which conclusions based on the impact analysis can be applied to similar programs, populations, and environments. Our feeling is that the generalizability is affected by the issues of sample selection and a number of current conditions.

1. Sample Selection

The primary issue here is representativeness of the sample (we'll accept as a given that large sample sizes are preferable). The sample is composed of members obtained from a larger population. Rarely, if ever, are measurements made on all members of a population. Analysts assume, however, that the measurements made on members of the sample are representative of measurements that could be made on the larger population to which the subjects in the sample belong. Therefore, if at all possible, one should employ random selection to ensure that the groups are the same before intervention and thereby preclude biased or systematic initial differences. If random selection is not employed, it should be so stated.

2. Current Conditions

The authors are indebted to Fairweather and Davidson (1986), who have sensitized us to a number of current conditions that one should evaluate before generalizing to another group or environment. Evaluating those conditions involves reviewing the study and answering three questions related to the *current* representativeness of the sample, the social context in which the model operated, and the outcome criterian used.

a. Representativeness of the Sample. The question here is, Is the sample used in the study *still* representative of the general population? A number of factors should be considered. One is the demographic characteristics of the program participants from which one wishes to generalize. If, for example, the intervention involved creating an innovative education program on junior high students, the generalization should be restricted to a similar group. Other factors to consider before generalizing include how participants entered the program (volunteer versus nonvolunteer, which adds a motivational component), differential success rates (for example, the program might have been more successful with some participants than others, and hence generalizations should be limited), and differential failure rates.

b. Social Context of the Model. We have repeatedly stressed that the context and environment of a program should be understood and clearly described (as we proposed in Section II on process analysis). We also have stressed that they constantly change. Thus, the representativeness of the social context within which programs function also requires reevaluation before generalizing. Factors such as organizational size, structure and climate, community acceptance, and political–idealogical climate are critical factors that may have changed and therefore will affect any generalization one attempts.

c. Outcome Criteria. The relevance of particular outcomes can also change and thereby reduce the representativeness of the outcome measures used. Reference was made previously, for example, to a sheltered workshop that was at the cutting edge in the 1960s but is currently being criticized because the criterion of social relevance has changed. In addition, the situational specificity of outcomes must also be considered, since many criteria come from archival measures, self-reports, and reports from significant others. Differences in these sources can cause problems in generalizing from one program to another, since the criteria may be different in another program or environment. This is frequently the case in education, rehabilitation, corrections, and mental health, wherein the criteria for functionality and relevance differ across agencies and professional groups.

D. Conclusion

In summary, one should not overlook the importance of internal and external validity. It is necessary to demonstrate the internal validity of the intervention so that one can support the impact statements made. As we saw, there are a number of threats to internal validity including history, attrition, maturation, regression, and measurement problems. The two primary issues involved in external validity (generalizability) relate to sample selection and a number of current conditions that need to be evaluated before generalizing to another group or environment. Our attempt has been to sensitize you to the importance of internal validity and generalizability, whether you are an evaluation producer or consumer.

VI. Summary

Estimating program impacts is not easy, because one needs to measure both sides of the structured comparison. In the simplest and most ideal sense, we estimate the impacts of a program by comparing the outcome measures for a group of participants with those for a comparison group. The challenge facing both the evaluation producer and consumer is to determine whether the program's outcomes were measured consistently under each situation or condition.

Our intent in the chapter was to focus primarily on the consumer who is reading someone else's impact analysis. In that regard, we discussed a number of procedures that can be used to estimate the effects of a program, a proposed impact analysis model, and three validity issues surrounding impact analysis. For the producer, we outlined two practical impact analysis techniques. Throughout the chapter, we presented a number of guidelines and cautions. Our guidelines included:

- The general procedure for estimating the impacts of a program is to compare the outcome measures for those who receive the intervention with the measures for those who do not.
- The certainty and precision of impact statements depend upon the comparison technique used. An experimental design results in the greatest certainty and precision.
- When you make assumptions in your impact analysis, and there is uncertainty in regard to those assumptions, sensitivity tests should be conducted to get a handle on the implications of the assumption(s).

We also had three cautions. First, if you use the impact analysis model outlined in Figure 8.3, don't forget that other things are occurring that are not reflected in the simple model. Second, it is okay to use the model for exposition and as a guide; but remember that it is weakened due to factors that are unmeasured and the assumptions made. And third, we cautioned you to remember that performance indicators, such as placement rates, are a necessary but insufficient condition in impact analysis.

Impact analysis addresses a number of critical evaluation questions such as (1) did the program have the intended effects on outcomes, (2) how big are the effects, (3) how much uncertainty is there surrounding the estimates, and (4) can these effects be attributed with reasonable certainty to the intervention being studied? These are important questions not just for impact analysis; they also provide the basis for benefit–cost analysis, which we discuss in the next chapter.

VII. Additional Readings

Boruch, R. F., & Rindskopf, D. (1977). On randomized experiments, approximation to experiments, and data analysis. In L. Rutman (Ed.), *Evaluation research methods: A basic guide* (pp. 143–176). Beverly Hills, CA: Sage.

Betsey, C. L., Hollister, R. S., Jr., & Papageorgious, M. R. (Eds.). (1985). *Youth employment and training program: The YEDPA years.* Washington, DC: National Academy Press.

Campbell, D. T., & Stanley, J. C. (1966). *Experimental and quasi-experimental designs for research.* Chicago: Rand McNally.

Collingnon, F. C., & Schmidt, B. (1978). *Measuring and valuing independent living impacts: A methodological review.* Berkeley: Berkeley Planning Associates.

Cook, T. J., & Scioli, F. P., Jr. (1975). Impact analysis in public policy research. In K. M. Dolbeare (Ed.), *Public policy evaluation* (pp. 95–118). Beverly Hills, CA: Sage.

Greenberg, D. A., & Robins, P. K. (1986). The changing role of social experiments in policy analysis. *Journal of Policy Analysis and Management, 5*(2), 340–362.

Haveman, R. H., & Weisbrod, T. (1977). Defining benefits of public programs: Some guidance for policy analysts. In R. H. Haveman & J. Margolis (Eds.), *Public expenditure and policy analysis* (2nd ed.) (p. 135). Chicago: Rand McNally.

Heal, L. W. (1985). Methodology for community integration research. In R. H. Bruininks & K. C. Lakin (Eds.), *Living and learning in the least restrictive environment* (pp. 199–224). Baltimore: Paul H. Brookes Publishing.

Sage Publications Series on Quantitative Application in the Social Sciences. (1978–1985). Beverly Hills, CA: Sage.

Weis, C. H. (1972). *Evaluation research: Methods for assessing program effectiveness.* Englewood Cliffs, NJ: Prentice-Hall.

IV

Benefit–Cost Analysis

Benefit–cost analysis is a process for weighing a program's benefits and costs. The process relies heavily upon the costing methodology outlined in Chapter 6 and the estimation of program impacts discussed in Chapter 8. Indeed, one cannot do benefit–cost analysis without information about costs and impacts.

Human service programs use a wide variety of resources to produce an equally wide range of outcomes. The inputs to this process include staff, buildings, materials, administrative resources, and support services. The outcomes include increases in a person's independence, productivity, and overall well-being. The primary issue addressed by benefit–cost analysis is whether the impacts of the program are big enough to justify the costs needed to produce them.

The extensive changes in human service programs during the 1970s and 1980s have created a dilemma for those conducting benefit–cost analyses. On the one hand, interest in this analytical method has increased dramatically as legislators, administrators, and consumers search for cost-effective programs in the face of growing demands for service and funding constraints at all levels of government. On the other hand, there is a concern that benefit–cost analysis, with its emphasis on dollars and cents, will fail to capture the humanitarian aspects of human service programs. The dilemma is how to conduct benefit–cost analyses that will retain the important summative features of this approach and, at the same time, capture all essential benefits and costs of the program.

As administrators and society make decisions about human service policy and programs, they seek to find programs that are efficient and equitable. Efficient programs are those that serve to increase the net value of the goods and services available to society. Equitable programs contribute to balancing the needs and desires of the various groups in society. Benefit–cost analysis is a tool developed for analyzing this type of situation. It is designed to facilitate assessments about whether a program produces effects that in some sense justify the costs incurred to operate the program. It does this by providing a means for organizing and summarizing information about a program so that it can be used to assess whether a program is likely to be efficient and to determine the nature of its effects on the distribution of income and opportunities. Benefit–cost analysis summarizes information about a program by attempting to measure in dollars all program inputs and all resulting effects. This enables the analyst to then compare all these aspects of the program directly by simply summing up the dollar values of all benefits and costs.

This approach is particularly well suited to helping analysts identify programs that are efficient. In fact, most early benefit–cost studies focused almost exclusively on efficiency. They sought to determine whether a given program was likely to increase the value of social resources or whether the resources that had been used for the program would have been better spent elsewhere. The suitability of this approach for assessing efficiency stems from its focus on resource use and on the dollar value of any resources that a program might use, save, or create.

In addition, benefit–cost analysis can also examine which groups in society will gain from a program and which groups will pay. This examination of equity is particularly important for assessing social programs, since a goal of many such programs is to increase social equity by reallocating resources or equalizing opportunities. In fact, for many social programs equity concerns dominate efficiency concerns since, other things equal, it is desirable to achieve a given equity goal as efficiently as possible.

While benefit–cost analysis seems quite useful for summarizing information and addressing both efficiency and equity issues, it has been adopted slowly by persons evaluating human service programs. In part, this reflects a sense that the emphasis benefit–cost analysis has traditionally placed on efficiency rather than equity is inappropriate for most human service programs. Other reasons for its slow adoption include a lack of useful paradigms for conducting benefit–cost analysis, considerable controversy involved in estimating dollar values for program effects, methodological problems involved in incorporating intangible effects that are often a central concern of human service programs, and the investment of time and resources needed to complete a thorough benefit–cost analysis.

This section of the book responds to these difficulties by presenting a benefit–cost analysis model for human service programs. The model provides a means for organizing data on program effects and costs to facilitate their use in policy discussions and decision making. The analysis recognizes the humanitarian aspects of human service programs while also assessing the economic efficiency of particular programs. It takes into account that from the start many benefits are difficult to appraise and quantify precisely.

The benefit–cost analysis model we present proposes that one use a standard procedure for valuing as many effects and costs as possible. The model thus provides a convenient and useful basis for comparing multiple program effects with each other and with program costs, for evaluating qualitative benefits and costs, and for comparing one program's benefits and costs with those of other programs. The model draws on several benefit–cost evaluations performed on employment and training programs, including the Job Corps (Long *et al.*, 1981; Thornton, Long, & Maller, 1982), the National Supported Work demonstration (Kemper, Long, & Thornton, 1981), the Structured Training and Employment Transitional Services (STETS) demonstration (Kerachsky *et al.*, 1985), and the National Long-Term Care demonstration (Thornton & Dunstan, 1986).

In addition, the model is based on the premise that you need to look at *all* benefits and costs of a program, even though you may be able to measure only some; conversely, some that are measured cannot be valued—but at least you can think about them. Thus, our benefit–cost analysis model is different from a cost-effectiveness model such as that proposed by Warner and Luce (1982). Although the costs are computed the same way in both models, the cost-effectiveness model involves selecting one outcome measure that is a good indicator of what you are trying to achieve, and then using the cost per unit of that indicator to compare programs. For example, you could look across five different programs and compare costs per placement, per months of residential program, per successful closure or treatment, or per dollar of postprogram wages earned. The major drawback of this cost-effectiveness approach, in our view, is that the analysis becomes meaningless if this indicator is not valid, or if it is defined too narrowly. We feel that the advantages of looking at all benefits and costs of a program are that it not only leads to a more complete analysis, but also minimizes the tendency to view benefit–cost analysis as a simple ratio of benefit to cost. We caution the reader about this reductionistic tendency, for it is important to realize that benefits and costs are multidimensional and that the total benefit–cost mosaic should be analyzed before making definitive statements or decisions.

Even though benefit–cost analysis is a powerful tool for evaluating the benefits and costs of human service programs, it includes numerous assumptions and estimates regarding costs and impacts. Therefore, there is considerable uncertainty involved in any benefit–cost analysis, and so we offer this caution:

Caution.   If benefit–cost analysis is not done carefully and accurately, it can easily be misused.

The benefit–cost model presented in this section provides a method for dealing with much of the uncertainty that surrounds program analysis. It does not eliminate the uncertainty that is inherent in evaluations of social programs; instead, it offers a way to organize the available information to facilitate decision making. The approach provides methods for judging the relative importance of different impacts and the importance of the uncertainty surrounding the estimates of those impacts. In fact, the approach emphasizes the value of benefit–cost analysis as a *process*. The real value of the technique lies in the process of systematically sorting through the available evidence rather than in relying on any single estimate of net present value or a benefit–cost ratio.

Benefit–cost analysis, just as impact analysis, is difficult because of what the analysis requires: information on costs and impacts. Indeed, you cannot do benefit–cost analysis without cost and impact statements. Thus, it can easily require resources either that are beyond your capability or that will throw off your resource balance. Our focus will therefore be primarily on your role as an evaluation consumer, although we do present strategies in Chapter 10 that allow you to use broad brush techniques to understand and assess potential program performance. The broad brush approach presented in Chapter 10 will also help you to sort through at the conceptual level the logic of the setup and rationale presented in Chapter 3 within a benefit–cost framework—factors such as who gains and who loses, how different groups experience the impacts, and what are the different perspectives on benefits and costs.

The section is divided into two chapters. In Chapter 9 we outline the procedures involved in the proposed benefit–cost analysis model, including examples from special education programs and the STETS demonstration project. In reviewing these benefit–cost methods, we realize that they typically imply a level of commitment beyond most programs. Therefore, we focus on specific problems in interpreting or addressing benefit–cost analysis across programs or evaluations. In Chapter 10, which is written for the producer, we discuss how one might do benefit–cost analysis on a limited budget or, as we propose, "on the back of an envelope."

In the end, benefit–cost analysis can be a helpful and powerful technique to improve decision making. But it requires resources, time, and expertise that are typically beyond most human service programs. Thus, funders should not expect programs to do full benefit–cost analyses, unless the program has the capability and is funded to do so. But if you have thought through the issues and applied the broad brush techniques we discuss, we feel that you will improve your program's performance and be in a better position to convince others that you have weighed your program's costs and benefits and, hence, its efficiency and equitableness.

9

Our Approach to Benefit–Cost Analysis

I. Overview

Benefit–cost analysis is essentially a structured comparison. Thus, as the first step in a benefit–cost analysis, the analyst must specify the program or policy being evaluated and the program or option with which it will be compared. This specification should include information on such factors as the persons being served, the treatments being offered, and the environment in which the program or policy will operate. These two alternatives—the program and the comparison situation—essentially define the scope, and ultimately the results, of the analysis. All work in the study, including the interpretation of the findings, must be undertaken in relation to these two alternatives.

As noted earlier, benefit–cost analysis makes the comparison between the two alternatives using the criteria of economic efficiency and equity. Specifically, it asks whether the decision to fund the program or policy under study will increase the aggregate value of social resources and whether it will produce desirable effects on the distribution of those resources compared to what would have happened under the alternative program or policy. The basic technique used to determine economic efficiency is to identify all changes in resource use caused by the decision to fund a program and then assign dollar values to those changes. The changes in resource use include those required to operate the program and those that result from the operations. The values of these changes are then summed together to yield an estimate of the program's *net present value,* the difference between the benefits and costs, where the dollar values of any benefits or costs that occur in future years are adjusted (that is, discounted) to reflect their value in a specified base period. A positive net present value indicates that the resources are being used more efficiently than they would have been under the comparison situation. A negative net present value indicates that (at least at its current scale) the program's resources could have been used more efficiently elsewhere.

The net present value criterion is also used to address equity issues. However, instead of aggregating all changes in resource use, the analysis considers the changes from the perspectives of the various groups in society that are affected by the program. For example, consider the students enrolled in a special education program. Part of the analysis of equity is to ascertain whether this group will benefit from their participation in the program. Similarly, the analysis will consider whether the taxpayers who fund the program obtain benefits that will outweigh the costs that they must bear. While the analysis can typically identify the major benefits and costs for these groups, it has no special criteria for assessing whether net shifts in resources between these and other groups are desirable. The appropriate criteria will vary, depending on the program under study and the groups affected. Thus, the value of shifts between groups must be determined within the broader context of public policy.

To estimate the value of changes in resource use, whether from the aggregate perspective of the economy or from the perspectives of particular groups in society, it is necessary to have a consistent means of assigning values to changes. The proposed model uses an approach based on the concepts that underlie the calculation of the gross national product (GNP). The GNP is a measure of the value of the goods and services a country produces in a year. Nonproductive activities, such as shifts in funds within or between sectors of the economy with no corresponding contribution to overall production, are excluded from the GNP calculation. This measure is estimated by aggregating the dollar values of all the goods and services produced, where the dollar values are the market prices of the various items being produced.

This general approach to accounting for economic production is also used in the proposed benefit–cost analysis model. The net changes in resource use attributable to a program or policy are first identified and then valued using market prices. As was the case for GNP, when a program merely reallocates resources (for example, by transferring them from one group to another) there is no net change in the aggregate resources available to society, and the shift in resource ownership will be excluded from the calculations. The advantage of using this approach is that market prices are readily observable. These prices thus provide a consistent and straightforward means for valuing changes in resource use. They also constitute a reasonable set of values, since market prices, which reflect the interaction of supply and demand, are generally viewed as good indicators of the relative values society places on different goods and services.

The strightforwardness of this approach is offset somewhat by the limitations of GNP calculations. GNP is the sum of all market activities and consequently excludes many activities that affect our well-being. For example, GNP excludes goods and services that are not sold in the marketplace, such as uncompensated work performed in the home by family members, child-care services provided by parents, and underground or illegal activities. Furthermore, the market values used in the GNP estimate may fail to capture true resource costs or social demands. For example, effects not captured by normal markets, such as environmental pollution emanating from production processes, are excluded from the GNP estimation. In addition, the social value of public services such as education or welfare programs may be inadequately captured from the amount of dollars spent to provide the services, and they are therefore inadequately represented in the estimated GNP.

Similarly, social programs often generate outcomes that have no observable market value, or the value of which is not adequately reflected in the interactions of the marketplace. These outcomes include the protection of individual rights and liberties, influences on public attitudes and perceptions, and the provision of special opportunities to specific needy populations. In any of these instances, the GNP-based approach is likely to exclude some key benefits or costs. These excluded items must usually be treated as unmeasured benefits or costs in the analysis, or an alternative means for valuing them must be proposed, such as using results from public opinion polls for valuing some outputs of social programs. Because of this problem, we offer the following caution:

Caution. Measuring and valuing the diverse inputs to social programs and the various outcomes is inexact.

This discussion indicates that while the net present value criterion is a fairly straightforward concept and can easily be defined, its actual application in benefit–cost analysis is difficult, since it requires that analysts make numerous assumptions and draw on many estimates. They must decide the appropriate dollar values to assign to program effects, make judgments about the correct interest and inflation rates for the period covered by the analysis, and assess the potential implications of any benefits or costs that could not be explicitly included in the analysis. All these assumptions and decisions make any estimate of net present value uncertain, and so any benefit–cost conclusions based on an application of the net present value criterion will also be somewhat uncertain.

To help meet the challenges posed by these problems and the dilemma of using benefit–cost analysis to study human service programs, we present in this chapter a benefit–cost approach that has been used successfully to evaluate a number of social programs. This approach emphasizes several features of benefit–cost analysis that make it particularly appropriate for assessing alternative program options, including:

- Use of a comprehensive accounting framework that includes all major benefits and costs, regardless of whether they can be explicitly measured or valued.

- Emphasis on benefit–cost analysis as a process rather than a bottom line—the knowledge gained by systematically assessing the available information about a program is generally more important than any single estimate of benefits and costs.
- The use of sensitivity tests to assess the relative importance and implications of the various assumptions and estimates used in the analysis.
- Multiple analytical perspectives that indicate how different groups in society will perceive a specific program and how the program will affect the distribution of social resources.
- A general approach to valuing program effects and incorporating unmeasured effects so that all essential effects can be taken into account when making decisions.

The chapter presents the analytical process of a benefit–cost study. In doing so, it defines and discusses each of the logical steps comprising this process, presents guidelines for establishing the analytical scope and framework of the analysis, provides standard procedures for estimating and valuing the full range of outcomes, and recommends how the results should be presented and interpreted.

The chapter contains four sections, the first three in which we use special education programs as our example. The first section discusses the procedures for setting up the analysis: defining the program and the comparison against which costs and benefits are measured; identifying the analytical perspectives of interest; and listing the expected benefits and costs. The second section examines the tasks needed to estimate the benefits and costs of a program: estimating the impacts of the program and their effect on resource use; assigning dollar values to the estimated impacts; and aggregating the benefits and costs that are valued. The third section addresses two major issues that are important in terms of presenting and interpreting the results: incorporating intangible effects in the analysis and undertaking sensitivity tests. And the fourth section presents a detailed benefit–cost analysis example from the STETS demonstration project.

II. Develop the Accounting Framework

The major steps involved in conducting a benefit–cost analysis are summarized in Table 9.1. As seen in the table, the first step is to define the scope of the study. What is the program being studied? What are the alternatives to that program? Who are the primary groups that will be affected by the program? What are the anticipated effects, and what resources will be needed to deliver the services? These are the questions that must be answered at the outset of the study. These questions pertain as much to a general assessment of a program,

Table 9.1. **Critical Steps Involved in Conducting a Benefit–Cost Analysis**

A. Develop the accounting framework.
   1. Define the program and its alternative.
      a. Program's objectives, clientele, services, operation, and environment
      b. Comparison situation (status quo; alternative program)
   2. Define the analytical perspectives.
      a. Social (society)
      b. Participant
      c. Nonparticipants (rest-of-society)
   3. List the benefits and costs (benefit–cost matrix table).
B. Estimate benefits and costs.
   1. Estimate the impacts of the program and their effect on resource use.
   2. Value the impacts of the program.
   3. Include intangible benefits or costs.
   4. Aggregate the valued benefits and costs.
C. Present and interpret the results.
   1. Include a benchmark net present value estimate.
   2. Include nonvalued impacts.
   3. Include a set of alternative estimates based on sensitivity tests.

---

such as we discussed in the section on process analysis, as they do to a benefit–cost analysis. They deal with the goals of the program and the means used to achieve those goals, which are issues of concern to all administrators. Thus, the first steps of the benefit–cost analysis have broad applicability. In many cases, sorting through these issues can be helpful for understanding a program, even if no further analysis and estimation is undertaken.

A. Define the Program and Its Alternative

The first task in a benefit–cost analysis is to specify the actual analytical comparison. The analyst must determine and define both the specific program or policy (or component of a program) being assessed and the alternative with which it is to be compared.

1. The Program

The important aspects of a program to be identified include its objectives, its clientele, the services it provides, how it operates, and the environment in which it operates. A careful enumeration of these features is a key aspect of the analysis, as we discussed previously in Section II on process analysis. It provides readers with an understanding of the program or policy being evaluated, and it is an essential first step for establishing the comparison under evaluation.

For special education programs, which we use to demonstrate the model's components, this program description is particularly crucial, since these programs are multidimensional and vary considerably. These programs typically encompass numerous policy goals and reflect an assortment of educational and therapeutic services provided to a disparate group of participants. The diversity of special education is nowhere more evident than in its goals. These include many social and psychological benefits, such as social integration, a more-educated society, greater self-esteem for the individual, and equal access and the right to educational opportunities. They also include economic benefits, such as enabling handicapped individuals to become more self-sufficient and socially productive.

Diversity is also evident in virtually all aspects of special education. The students differ substantially in their abilities and special needs. There are also differences among districts with regard to the ages of the students entitled to services, which range from preschool children to high school graduates. There are differences in the types of instructional and therapeutic services offered; facilities, equipment, and personnel used; and educational settings. Finally, there are differences in the local environments that will affect program operations. These include transportation services, the supply of trained teachers, the local tax base, and the types of residential facilities available to persons with handicaps.

With all this variation, it is essential to define exactly what is being studied. An analyst might consider all special education programs or particular individuals or services. There may be interest in, for example, a specific subset of students (for example, preschool children with mental retardation or blind high school students), a specific type of service (for example, early intervention or physical therapy), a specific educational placement (for example, special residential schools or regular classrooms), or specific political subdivisions (for example, how different local districts implement a given state policy).

As an example of this process, consider a local school district that wants to assess the desirability of funding a new vocational education program for special education high school students. The definition of this program must be sufficiently detailed that the analyst and user of the study will know exactly what is being examined. Such a definition should include the program goals, its expected effects, the students who will be served (including their backgrounds and the manner in which they will be selected), the services and curriculum offered, and the environment in which the services will be offered. The description of the environment is particularly important, since it can have a strong effect on the results of the study. For example, the observed success of a vocational education program with the goal of increasing employment for high school graduates will depend on the nature of the local labor market and the availability of adult services to assist graduating students in their job search and to help them remain on that job.

2. Comparison Situation

The relevant comparison situation depends upon the issues of interest to policymakers and the environment in which a program operates. Ultimately, however, the choice of the comparison situation is a critical decision to the analysis, since this decision defines the analytical comparisons and drives all of the results that are generated. Consequently, defining the comparison situation in clear, exact terms is an important analytical task, as we saw in the preceding section on impact analysis.

The comparison situation can take several forms. It may be the status quo, defined as the mix of currently available services and programs (if any). Or it may be a specific alternative program or policy. In our illustrative example in the preceding section (a new vocational education program intended specifically for special education high school students), we will assume that the comparison situation is the services available under the status quo. Specifically, we will assume that in the absence of the new vocational services designed for special education students, such students would receive only nonvocational education. Some students might enroll in adult service programs outside the school system, but such enrollments would not be specifically linked to their education or school.

Defining the comparison situation ultimately requires a detailed examination of the services that those students who are enrolled in the new vocational program would have received either in the absence of the new vocational program or in an alternative program, depending upon the alternative chosen. Again, for either comparison situation, the nature and objectives of available treatments (if any), the operation and integration of these programs, and resource needs and uses must all be defined clearly. Thus, our first guideline regarding benefit–cost analysis:

Guideline 29. The first step in conducting a benefit–cost analysis involves defining the program and its alternative.

B. Define the Analytical Perspectives

Any public policy or program typically will affect many groups of individuals. For example, a special education program will clearly affect participating students and their families and may have long-run effects on special agencies and employers in the community. It will also have an impact on government budgets and hence indirectly will have an effect on taxpayers. Each of these groups has a perspective on the program, and each of these perspectives will have some relevancy to decision making.

For the issue of economic efficiency, the perspective of society as a whole is relevant. It captures the net effect of the program on the aggregate value of available goods and services. Equity issues are addressed through the perspectives of specific groups affected by the program. These other perspectives are just as important as that of society as a whole, given the nature of how decisions are made in our society. They indicate the incentives different groups have to support or oppose a program.

The perspective of participating students is particularly important, since it indicates the extent to which they will benefit from the education services intended to help them. It is also useful to consider the perspective of everyone else in society—that is, all the persons not enrolled as students in the program. This perspective will capture all effects that do not accrue to the students. In particular, it will capture the taxes needed to finance the program and any resulting reductions in expenditures for alternative programs.

Depending upon the particular equity concerns of the analysts, other perspectives may also be used. For example, in an analysis in which the comparison situation is an alternative program that affects a different set of participants, it may be useful to illustrate the effects of the policy under analysis from the perspectives of the two sets of potential participants. Alternatively, an analyst may find it useful to disaggregate the rest-of-society perspective into funding sources (for example, federal, state, and local taxpayers). However, since the analysis will require estimates of effects on each perspective, the complexity of the analysis increases rapidly with the number of perspectives, and it is usually preferable to disaggregate society into only the two major perspectives described—participants and the rest of society.

It is important that, when defining or selecting perspectives for an analysis of equity, the groups chosen be mutually exclusive yet, in total, include everyone in society. By defining perspectives in this way, the sum of the valued benefits and costs for each individual perspective will equal the net effect as seen by the perspective of society as a whole. In our framework, the sum of benefits and costs accruing to program participants (for example, participating students) and the rest of society will equal the net benefit to society as a whole. It should be noted, however, that this "adding-up" property necessitates assuming that a dollar of benefit or cost to one person is equivalent to a dollar of benefit or cost to any other person. The perspective of society as a whole would thereby ignore all redistributional questions and focus on aggregate resource use questions. This is a convenient, but not a critical, assumption. The analysis could assume other distributional value systems, giving more or less weight to the resources owed by specific groups in society. However, given the difficulty of defining and using such a system, as well as its inherently controversial nature, we recommend that the "equal value" system used here be adopted.

This relationship of analytical perspectives provides a convenient structure to the analysis. The perspective of society as a whole focuses on economic efficiency—that is, the change in the total resources of society caused by the program under evaluation (relative to the alternative comparison)—and ignores all redistributional aspects of the program. Because the participant and rest-of-society perspectives represent mutually exclusive groups that in total represent all of society, only those benefits (or costs) that accure to one group within society and have no equal offsetting cost (or benefit) to the other group (that is, those outcomes that involve the use or production, rather than the redistribution of resources) will remain in the social perspective. Others will net out to zero in the social perspective. Consequently, the social net present value of calculation will include only the value of outcomes that affect the total amount of resources (goods and services) available to society. The equity implications of the program, on the other hand, can then be assessed by examining any net shifts in resources between participants and the rest of society. Focusing on the different analytical perspectives leads to our next guideline:

> *Guideline 30.*   The three most appropriate analytical perspectives are social (society), participant, and the rest of society (nonparticipant).

C. List the Benefits and Costs

A list of the expected benefits and costs is the next step in developing the accounting framework. This step follows directly from defining the program and the analytical perspectives. The results of this step will define the specific data the analyst ultimately will need in order to estimate benefits and costs. Therefore, the results of this task are an important and necessary input into designing the actual analysis and estimation of impacts described in the next section. Its importance is reflected in the following guideline:

> *Guideline 31.*   The comprehensive accounting framework should include all benefits and costs, regardless of whether they can be measured or valued.

At first consideration, this guideline appears relatively straightforward. Yet when one actually gets involved in attempting to identify and sort out the various types of potential effects, particularly from the different viewpoints of the relevant groups affected, it becomes evident that this process requires careful consideration of the interactions of outcomes among these different perspectives. Thus, before presenting a suggested benefit–cost matrix table, we first discuss two general rules that the analyst must keep in mind when developing the list of expected benefits and costs for the accounting framework.

The first general rule is that the analyst must try to be as comprehensive as possible when considering the expected impacts and resource uses of a program, even though not all expected benefits and costs will actually be valued. The analyst must attempt to identify all changes in behavior or outcomes that would lead to a real change in the use or availability of resources. For example, special education programs may produce a change in students' use of a variety of other support service programs. Since a change in the use of such services represents a real change in the use of resources available to society, the effect of this change in behavior should also be included in the benefit–cost calculation. It is also important that the benefits and costs that cannot actually be valued monetarily be identified and accounted for in the framework. Sometimes, the programmatic impacts and desired outcomes that cannot be valued are those that are most important—that is, those that tip the scale toward making a positive assessment of a program when the result of the net present value calculation for those outcomes that are valued monetarily is zero or less.

Given the likely constraints on the scope of an analysis, the analyst will have to set limits on this process. For instance, research (see Haveman & Wolfe, 1984; Hanushek, 1986) provides evidence of a relationship between schooling and a variety of outcomes, including labor market performance, nonmarketed home production, health status, child-rearing, crime, social cohesion, marital and divorce rates, charitable donations, political socialization, and voting behavior. An exhaustive effort would be necessary to identify and measure all the secondary effects of special education. Consequently, the analyst, purposefully being comprehensive yet working under time and resource constraints and in recognition of the level of uncertainty associated with estimating the magnitude and value of certain effects, will have to set priorities. The analyst may set such priorities by identifying only those impacts that either represent effects that are expected to exceed a particular magnitude or can be attributed directly to the program under study with a reasonable level of certainty.

The second general rule in identifying the benefits and costs of a program is that benefits and costs are measured relative to what would have happened under the comparison situation. As we stated earlier, the net present value of a program summarizes a comparison between that program and another policy option or program. Each of the identified benefits or costs of a program represents the difference between the expected outcomes or resource of the policy being evaluated and those under the specified comparison situation. Therefore, benefits and costs are typically measured as changes or differences between what would have occurred under the comparison situation and what actually did occur under the intervention. For example, a vocational special education program may generate an increase in a student's lifetime earnings over what his or her earnings would have been in the absence of the program. Similarly, this program may reduce (relative to the comparison situation) the level at which certain ancillary services may have been used. The impacts to be measured as benefits in these examples are the increases in earnings or the reductions in ancillary-service use.

If a particular use of resources would be the same under either the program or the comparison situation, then it should be omitted from the framework. For example, consider the resources involved in transporting special education students to a vocational education program. If the alternative is to place them in a nonvocational program that would require the same amount of transportation, then the choice between alternatives would not affect transportation costs; thus, they could be excluded from the framework, despite the fact that transportation is an expensive cost item. In this way, the framework can be simplified, and analytical resources can be devoted to the critical changes produced by the program.

Thus, in developing the list of expected benefits and costs for the accounting framework, be sure to keep the following guideline in mind:

> *Guideline 32.* The two general rules to follow in developing the list of expected benefits and costs are the following: (1) be as comprehensive as possible when considering the expected impacts and resource uses of a program, and (2) measure benefits and costs relative to what would have happened under the comparison situation.

Table 9.2 presents a list of benefits and costs for our example in which a school district is evaluating funding a new vocational education program specifically intended for special education high school students. The comparison situation for our example is the status quo, in which there is no vocational education program specifically for such students who could use other available nonvocational special and regular education services and the available services from other providers.

Three perspectives are shown in Table 9.2: society as a whole and two specific groups that comprise society as a whole—students who are offered the vocational education program and the rest of society. How the anticipated impacts of the program are expected to be perceived (that is, as benefits or costs) by these different groups is also indicated. Untimately, the evaluation will empirically assess whether these expectations are accurate and whether the measured impacts actually represent benefits or costs.

The primary purpose of the benefit–cost matrix format presented as Table 9.2 is to help organize and conduct the analysis and ensure that all the major

Table 9.2. **Expected Benefits and Costs of a Hypothetical Vocational Educational Program for Special Education High School Students**

Impacts	Analytical perspective[a]		
	Social	Student	Rest-of-society
Benefits			
Increased output			
Increased employment	+	+	0
Indirect labor market effects	0	0	0
Increased taxes	0	−	+
Work preferences	+	+	+
Reduced use of alternative programs			
Alternative school programs (e.g., non-vocational education)	+	0	+
Job-training or work-related programs (e.g., work activity centers, sheltered workshops)	+	0	+
Supportive services (e.g., transportation, healthcare, counseling, housing)	+	0	+
Program allowance	0	−	+
Reduced use of transfer programs (SSI, welfare, food stamps, etc.)			
Reduced benefit payments	0	−	+
Reduced administrative costs	+	0	+
Other benefits			
Increased self-sufficiency	+	+	+
Increased self-esteem	+	+	+
Improved quality of life	+	+	+
Costs			
Program costs			
Operational costs	−	0	−
Administrative costs	−	0	−
Foregone nonmarket output	−	−	0
Increased use of complementary programs (e.g., transitional employment programs, employment service, supported employment programs)	−	0	−

[a]The individual components are characterized from the three perspectives as being a net benefit (+), a net cost (−), or neither (0).

impacts of the program are captured accurately in the analysis. The analytical process involved can be summarized as follows: for each outcome that the program is intended to produce (as determined from the program description), the analyst must (1) identify all changes in resource availability (relative to the comparison situation); (2) assess whether those changes actually create, save, or use resources or whether they simply redistribute resources among groups; and (3) determine how these changes will affect the various perspectives.

In Table 9.2, we have followed this procedure to complete a list of benefits and costs for the hypothetical vocational education program for special education students. We have included benefits from the expected increased employment of students and the reduced use of other education, training, employment, income maintenance, and social service programs. As costs, we have included the costs of program operations and any increased use of other programs that may be related to the program under study (for example, transitional employment or job-placement services provided to graduating students). We have also included as benefits a number of program effects that would be difficult to value in a benefit–cost analysis, including changes in self-sufficiency, self-esteem, quality of life, and nonmarket production.

For the most part, the process of filling in the table is a straightforward application of the guidelines and procedures described above. However, some outcomes do require particularly careful analytical reasoning to identify real resource changes and to sort out the effects of these changes from the various relevant perspectives. We discuss in more detail two specific program effects: increased employment and changes in the use of other programs or services.

1. Increased Employment

The employment outcomes of the special education students who enrolled in the new vocational education program will affect not only the students but also the other persons in society. The analysis must capture these effects as well as the effect on the overall level of resources available to society.

The students' perspective is fairly straightforward. They voluntarily accept the employment, and they decide to forgo other opportunities in order to accept that employment. The measure of their net gain is the difference between their actual employment and what their employment (or other activities) would have been under the comparison situation. In our example, we expect that a net increase in employment will occur as the vocationally trained students graduate and obtain on average more and better jobs, compared with the jobs that they would have obtained (if any) in the absence of the vocational education program. This increase in employment will benefit the students financially in the form of increased earnings and fringe benefits (although they will have to give up part of their increased earnings in the form of taxes). The students may also benefit beyond their pecuniary compensation, if they derive psychological benefits from their employment (for example, greater self-esteem). Both types of benefits enter into the accounting framework in Table 9.2.

The situation for the rest-of-society group is more complicated and depends on the mechanics of the labor markets in which the students obtain jobs. In general, the rest-of-society group benefits from any increased taxes paid by the student group, since these students will be paying a larger share of the total tax burden. However, two subgroups of the rest-of-society perspective are affected more directly: the employers of the vocational education students and the other workers who compete in the same labor market as those students.

For employers, it is probably a reasonable assumption that they break even—that is, that the wages they pay are equal to the value of the output of their workers. The equality between the cost of labor (that is, the compensation paid to workers, including both wages and fringe benefits) and the value of output is based on the assumption that product and labor markets function competitively. Union power, government wage subsidies, monopolies, taxes, and other factors can work to break this equality. In general, it is best to assume that the value of output is accurately measured by total compensation, unless there is specific evidence to the contrary—for example, if students were placed in subsidized jobs.

Since this resource interchange (employers gaining the value of output while paying the compensation) nets out to zero and is internal to the one perspective (that of the rest of society), this redistribution of resources can be excluded from the framework. However, if such resource uses are critical to the particular analysis question or if the equality between the cost of labor and the value of output does not hold, then a separate perspective may be established to keep track of the distributional implications.

For other workers, the situation depends on whether any indirect labor market effects exist (Hall, 1979; Johnson, 1979). Indirect labor market effects occur in the presence of unemployment. If the students take jobs in a labor market in which workers are unemployed (that is, in which an excess supply of labor exists), they take jobs that would have been filled otherwise. Alternatively, if students forgo jobs in such a market, those jobs would be filled by workers who would have been unemployed otherwise. The presence of these types of effects—termed *displacement* and *replacement,* respectively—depends on the types of jobs that the students obtain and those that they would have obtained under the comparison situation. The most favorable case for other workers would be one in which the vocational education program enabled students to avoid entering a labor market characterized by high unemployment and, instead, obtain jobs in a market characterized by an excess demand for labor. In this case, the other, unemployed workers in the market would gain, since they would fill the jobs that students forgo and, due to the shortage of labor in the market that students enter, no other workers would be displaced. The opposite case—students entering a market with an excess supply of labor—would impose costs on the other workers, as the students take jobs that otherwise would have been held by someone else.

It is clearly difficult to measure these effects. It requires information on the local labor markets, the jobs held by students, and the decisions of affected employers and workers. Consequently, indirect labor market effects are commonly ignored or assumed away. This approach may be problematic for many special education programs in which students are placed in entry-level, minimum-wage jobs. In the past, such labor markets often exhibited a considerable degree of unemployment, and the likelihood of displacement was high. Programs may have simply shuffled workers among a fixed number of jobs (substituting the graduating students for other low-wage workers), with no net increase in aggregate employment. Recently, however, this situation has improved, with signs that the relative supply of entry workers is declining (at least at prevailing wage rates).

In listing the benefits and costs in Table 9.2, we have included a line for indirect labor market effects. Nevertheless, we have continued to follow the conventional approach of assuming that the rest-of-society perspective is unaffected by any changes in students' employment, except for any preferences of society to have the students be more self-sufficient and integrated. This assumption reflects the lack of empirical evidence on indirect labor market effects.

A benefit–cost analysis should assess whether this assumption is reasonable, as well as the implications of changing it. If there is interest in the potential of a program under full employment (that is, no displacement or replacement), then the assumption is clearly reasonable and little additional analysis is required (see Kemper & Long, 1981). Conversely, if unemployment is assumed to persist, then an accurate program assessment necessitates examining the indirect labor market effects. This is generally done with sensitivity tests that estimate how the overall benefit–cost conclusions would change under alternative assumptions about the extent to which indirect labor market effects exist. We return to the issue of sensitivity tests later in the chapter.

Once the participant and rest-of-society perspectives have been resolved, they can be summed to yield the overall social perspective. In our example, increases in the value of output by the students (as measured by increases in their total compensation) are not offset by any indirect labor market effects on other workers; thus, increased output enters the social perspective as a real increase in the value of social goods and services. Any psychological benefits (in excess of the value of output) that students or others derive from the increased student employment will also enter the social perspective. Given a fixed total tax burden, the increased taxes paid by students net out against the increased tax receipts of the rest of society. That is, they represent a redistribution of resources rather than a resource use, and as such are excluded from the social perspective.

2. Changes in the Use of Other Programs

In addition to the effects on employment, special education programs (as with most human service programs) can be expected to influence the use of a wide range of other programs. These include income support, vocational, residential, transportation, and social service programs. In all of these cases, the special education program may directly affect the extent to which its students use these other programs and may indirectly affect the extent to which other persons use these programs.

III. Estimate Benefits and Costs

Thus far, we have described a general model for analyzing the benefits and costs of special education programs. We have focused on defining the analytical comparisons to be addressed and specified the various outcomes to be studied. We now turn to the methods of using that framework to assess whether the program or policy under study generates impacts that are, in some sense, worth the costs required to produce them.

In designing the empirical work, the analyst must determine how much precision is needed. The required level will be determined by the needs of policymakers and by the magnitude of the policy question under study. Many decisions involve the commitment of only a modest level of resources or are those that can easily be changed over time. These decisions can be analyzed with an informal weighing of benefits and costs. Other decisions involve committing substantial resources, since they require more precise information and a more rigorous comparison of benefits and costs.

The precision of the analysis will be determined by the evaluation design and the analytical methods used. The number of persons observed, the procedures for selecting those persons, data sources, data collection procedures, statistical techniques, and analytical methods all interact to produce the level of confidence and certainty that can be placed in the empirical findings. The challenge facing the benefit–cost analyst is to choose among these various factors in order to produce estimates that will provide a good basis for the decisions that must be made. In this regard, the analyst must be sensitive to how the analysis will be applied and to how the various trade-offs that are made in designing the evaluation will affect the consequent level of certainty. In general, the more certainty required by the policy audience, the more rigorous the evaluation must be.

The analyst must consider four general areas when designing the empirical components of a benefit–cost analysis (see Table 9.1). They are the following: (1) estimating the magnitude of program impacts, (2) valuing the impacts of the program, (3) including intangible benefits or costs, and (4) aggregating the valued benefits and costs.

When evaluating the effects a vocational education program might have on the use of other programs, it may initially appear that these indirect effects resemble the indirect labor market effects of displacement and replacement. For example, if special education students participating in our hypothetical vocational education program reduce their use of other employment and training programs, then new openings may be made available to students who otherwise would have been excluded. Alternatively, if the vocational education students increase their use of other programs that supplement their vocational education, they may fill positions that otherwise would have been filled by other students.

However, these indirect effects differ from those that occur in the labor market, and therefore their treatment in the benefit–cost analysis is different. The essential difference between the indirect labor market effects and those for other programs is that changes in the resources devoted to programs reflect government decisions about resource allocation rather than constraints on the economy. When students are diverted from other programs into the new vocational education program (as we assume in our example), the resources that would have been used to provide those other program services could be reallocated to alternative uses. One such alternative would be to reduce the size of the other programs by not filling the spaces vacated by the students who left to enter the vocational education program. In this case, the freed resources could be devoted to other social programs, used to pay for the new vocational education program, or used to reduce the overall government budget. Alternatively, the resources could be left with their original programs, enabling those programs to serve persons who otherwise would have been unserved. In any of these cases, the resources that are saved when the vocational education students reduce their use of alternative programs are benefits to the persons who would have had to pay for them (in most cases this group will be the taxpayers). This is true regardless of how those persons choose to reallocate and spend their savings.

This situation differs from the indirect labor market effects because unemployment is assumed to be involuntary. Unlike decisions about the resources to be devoted to programs, unemployment is viewed not as a conscious decision about the allocation of resources but, rather, as an unwanted condition resulting from constraints on the economy. There is a presumption that unemployed workers would like to work, if they could find jobs. If a worker is displaced from a job, there is a resource "saving" in the form of that worker's time. However, unless there is full employment, that worker will be able to reallocate that time to his or her preferred option, a new job. Thus, there is a loss in the aggregate value of social resources measured by the lost earnings of the displaced worker.

Given that changes in the use of alternative programs can be analyzed without worrying about indirect effects, the central issue facing the analyst is to determine which programs will be affected and whether those effects will be benefits or costs. This issue is best addressed by reviewing all of the programs that are used by the persons who are enrolled in the program under study. A change in school curriculum and any subsequent changes in adult activities are likely to affect the use of a wide range of programs, so this process should be as comprehensive as possible. It is essential to include programs that are substitutes for the program under study as well as those that are compliments. One would expect to reduce the use of substitute programs, thereby creating benefits, and to increase the use of complimentary programs, thereby incurring costs.

In Table 9.2, we have assumed that the vocational education program will substitute for other nonvocational school programs and will reduce the need for some adult vocational and support programs. At the same time, we have assumed that the vocational education students will make greater use of programs such as transitional employment, as they move from school to work. These are reasonable expectations, but changes in program use will ultimately be classified as benefits or costs depending on the actual use patterns observed in the evaluation.

The list of benefits and costs presented in Table 9.2 is only one example. While many special education programs will generate these benefits and costs, the list of benefits and costs will clearly be different for other programs, groups, or comparison situations (as we see later with the STETS example). Additionally, under an evaluation of an early intervention program (compared with no early intervention program), a significant benefit may be the cost savings associated with reducing the special education services which would otherwise be required later. Such cost savings would represent a benefit to the nonparticipant taxpayer (that is, the rest of society) and an increase in the resources available to society. Other benefits may be derived from the effects of an early intervention program to the extent that it may reduce dropout rates (indirectly affecting future output and all other educational costs and benefits) and generate psychological benefits by enabling students to participate as fully as possible in regular classes and shed any ''special education'' stigma sooner than they might have otherwise.

The key to this process is to start with precise definitions of the program or policy being evaluated, the comparison situation, and the perspectives of interest. When these elements are specified, listing the benefits and costs is usually straightforward, although some issues (such as the indirect labor market effects and changes in the use of alternative programs) require careful analysis. For additional applications of the model, the interested reader is referred to benefit–cost analyses of transitional employment for mentally retarded young adults (Kerachsky *et al.*, 1985), long-term care for the elderly (Thornton & Dunstan, 1986), employment and training programs (Long *et al.*, 1981), apprenticeship programs (Mallar & Thornton, 1980), offender rehabilitation programs (Maller & Thornton, 1978), and social programs generally (Thornton, 1984).

A. Estimate Program Impacts

Since Chapter 8 was devoted to estimating program impacts, we provide only a summary overview in this section. As stressed throughout Chapter 8, the most difficult task facing analysts is to determine the magnitude of a program's effects. The determination requires estimating the extent to which the activities of participants enrolled in the program differ from the activities those persons would have had in the comparison situation. As discussed in Chapter 8, there are four ways to estimate what a group of participants would have done under alternative conditions: experimental design, using a comparison group, using a hypothetical comparison group, or using pre–post measures on the same group. The reader might want to refer back to Figure 8.2 and Table 8.2 to review the relationships among these four comparison techniques, assumptions made, and the precision of each analysis. In reference to estimating programmatic impacts, we suggest the following guideline:

Guideline 33. Estimating programmatic impacts involves comparing the outcome measures for those who receive the intervention with the measures for those who do not. The certainty and precision of impact statements depend upon the comparison technique used; an experimental design results in the greatest certainty and precision.

Regardless of the comparison technique adopted, the data collection process will be a major issue to resolve, as we discussed in considerable detail in Chapters 4 and 7 (Table 4.2 and Table 7.3). Data sources for our special education example include the students, their parents, teachers, school records, and the records from other agencies that might be affected by the intervention (for example, state vocational rehabilitation or developmental disabilities departments, the Social Security Administration and state unemployment insurance systems). The collection procedures include interviews (either in-person, telephone, or mail), extracts from service providers, and existing data systems.

It is also essential that all data sets be consistent with each other, since the accounting framework necessitates that benefits and costs be measured for the same set of individuals. In particular, it is essential that the costs included in the accounting framework be those that produced the outcomes included in the framework. Thus, the analyst must ensure that the program costs included in the analysis accurately represent those actually incurred to serve the treatment group. To do so, the analyst usually follows the activities of a single group of individuals over time and monitors the costs of serving them. If costs and impacts are estimated based on the experience of different groups, then it is incumbent upon the analyst to document whether the two groups are essentially identical.

B. Value the Impacts of the Program

Once the impacts of a program have been measured, the analyst must determine the value of these effects. Two general methods are used to assign dollar values; the applicability of these methods depends on what the effects are and how they are measured. The first and simplest method is to use dollar-denominated outcome measures. The second method is to estimate the value of the effects based on estimated prices (termed shadow prices) that reflect the estimated value per unit of change in behavior or outcomes.

1. Dollar-Denominated Outcome Measures

The simplest method for estimating the value of an impact is to use, if available and appropriate, outcomes that can be denominated in dollar terms. In a benefit–cost analysis, a number of impacts should be measured directly in terms of dollars. For instance, in the example of the vocational education program for handicapped students, a researcher can attempt to measure the impacts of the program on earnings, transfer payment receipts, tax payments, and Medicare or Medicaid benefits. For such variables, the key issue is whether the data sources, data collection procedures, and estimation techniques will produce accurate estimates. It is also necessary to ensure that the dollar values have been adjusted for inflation, an issue we consider in the next section.

2. Shadow Prices

Shadow prices are essentially prices that are not determined by markets. Instead, they are estimated by the analyst to reflect the average resource value of specific activities or goods. These shadow prices are used in the analysis in a manner similar to how unit prices are used in regular markets—the change in an outcome is multiplied by the appropriate shadow price in order to estimate the total value of the change.

Shadow prices are a major source of uncertainty in benefit–cost analysis. In part, such uncertainty reflects the concern that since these prices are not determined by the workings of a competitive market they may fail to capture the true resource cost of changes in the outcomes of interest. The uncertainty also reflects problems associated with making the estimates, which range from conceptual problems about correctly defining a shadow price for an outcome, to estimation problems caused by inadequate data. Analysts should recognize this uncertainty and examine its implications by documenting the shadow price estimates to enable readers to form their own judgments about the appropriateness of these estimates. Analysts should also conduct sensitivity tests that indicate how the final net present value estimate would change in response to variations in estimates of key shadow prices. Such tests can indicate the degree to which uncertainty about a specific shadow price tempers the overall benefit–cost conclusions.

An example of the use of shadow prices can be shown by estimating changes in students' tax payments. It is anticipated that increases in earnings will lead to increased tax payments by participants. Without shadow prices, a researcher would have to collect data on changes in payroll taxes, local, state, and federal income taxes, and sale and excise taxes—a complex and expensive process. Alternatively, changes in tax payments can be estimated more easily by multiplying available estimates of effective tax rates by estimates of changes in income derived from the impact analysis. Pechman (1985) has estimated these effective tax rates (including all types of taxes) on total income for various income classes. Table 9.3 summarizes his findings.

A similar shadow-pricing effort is necessary for estimating impacts on fringe benefits, which must be included as part of total compensation when valuing employment effects. Fringe benefits include workers' compensation taxes, medical insurance contributions, pension and retirement contributions,

Table 9.3. **Effective Tax Rates by Adjusted Family Income under Alternative Incidence Assumptions (1980)**[a]

Adjusted family income (thousands of dollars)[b]	Taxes as a percentage of total income	
	Most progressive incidence assumptions[c]	Least progressive incidence assumptions[d]
0–5	32.5	57.7
5–10	20.3	26.7
10–15	20.5	24.9
15–20	21.5	25.0
20–25	22.7	25.6
25–30	23.2	25.9
30–50	24.5	26.6
50–100	26.5	26.9
100–500	27.3	24.1
500–1,000	27.0	19.8
1,000 and more	31.0	20.7
All classes (mean value)	25.2	26.3

[a]Effective tax rates are taxes paid as percentage of income. Taxes accounted for here include federal, state, and local income, corporate profits, excise, sales, and property taxes.
[b]Family income includes employer compensation, earnings on investments (for example, net rental income, dividends, capital gains), transfer payments, and nonmoney income (that is, value of food stamps, Medicare, and Medicaid; net imputed rent on owner-occupied dwellings; and unrealized gains).
[c]The most progressive incidence assumptions assume that both employee and employer payroll taxes are borne by employees in proportion to their level of earnings, and that corporate income and property taxes are distributed in proportion to reported property income.
[d]The least progressive incidence assumptions are that half of payroll taxes on employers and half of corporate income taxes are shifted to consumers.

and unemployment insurance. Paid vacation and sick leave are also fringe benefits, but these are best measured by simply including vacation time or sick-leave
time in the estimates of total time employed and total earnings. In this case,
estimates of fringe benefit rates can be multiplied by the estimated effects on
earnings to estimate the effect on fringe benefits.

Estimates of fringe benefit rates can be obtained from several sources. For
example, the U.S. Department of Labor (1980) examined average compensation
levels for low-wage workers (that is, workers who earned less than $3.00 of total
compensation per hour in 1977, which would be approximately equal to $5.50 an
hour in 1986 dollars) in the private nonfarm economy. They found that in 1977,
wages and salaries for these workers accounted for 85% of total compensation.
The remaining 15% of employer payments went for various fringe benefits.
These figures indicate that fringe benefits would represent, on average, almost
18% of total wages and salaries (that is, 0.15 divided by 0.85). This method is
useful for estimating fringe benefits for the target populations of many social
programs, since the estimate covers a very wide range of jobs and reflects the
position of low-wage workers who are often the target of such programs. However, it is based on relatively old data and may fail to capture recent changes in
mandated fringe benefits such as Social Security taxes, workers' compensation,
and unemployment insurance.

The Bureau of Labor Statistics no longer collects the data that were used to
estimate this fringe benefit rate. Other sources are available, but they are somewhat limited by the types of firms that they study and the range of high- and low-
paying jobs that are included. However, they generally indicate fringe benefit
rates similar to those from the earlier Labor Department study. For example,
Jones (1986) examined data from the 1984 Survey of Current Business and
estimated that average fringe benefits (excluding those for time not worked such
as vacation and sick leave) were 21%. Also, the U.S. Chamber of Commerce
(1986) analyzed data from 1,000 firms and estimated that fringe benefit rates
average 25% (again, excluding payments for time not worked).

In practice, it is probably best to select an estimate from the 18–25% range,
and then test the importance of that selection by conducting sensitivity tests.
Rates in the lower part of this range should be used when participants in the
program being evaluated will work predominately in low-wage jobs. In any
event, the results of benefit–cost analyses of social programs are unlikely to
depend crucially on the estimated fringe benefit rate, and so the use of any
specific rate from this range is likely to be acceptable.

Shadow prices can also be estimated for special education programs. Lewis,
Bruininks, and Thurlow (1986) estimated the average annual costs per handicapped student for 12 program service areas (for example, occupational therapy,
adapted physical education, and hearing impaired) for an early intervention
program and in total for the special education students served during the 1983–

1984 school year in one school district in Minnesota. A study by the Rand Corporation (Kakalik, Furry, & Thomas, 1981) estimated the annual nationwide costs per handicapped student of special education programs by type of educational placement, type of instruction, age, and type of handicap during the 1977–1978 school year. These studies indicate that program costs vary considerably by type of service, client, handicapping condition, educational placement, and type of instruction. In addition, local variations in prices or costs for services and goods will influence cost estimates. Thus, cost estimates made for one program or one time period may be inappropriate for other programs and time periods.

Such variation, which affects virtually all shadow-pricing efforts, implies that analysts should examine program definitions carefully to ensure that appropriate shadow prices are used. This is particularly true when the shadow prices are used to estimate costs that play a central role in the evaluation, as would be the case for the costs of the intervention being studied.

Estimates of the average benefit payments and average administrative costs of the various income transfer programs, as well as estimates of the average costs of various types of employment, education, counseling, recreational, housing, or transportation programs, may be used as shadow prices to represent the value of the resources used in the programs for a given unit of time. Although the ideal situation is to use marginal costs (that is, the change in total program costs caused by the addition or subtraction of one participant), average costs typically are used because they are easier to measure accurately and, under plausible assumptions, should be quite close to long-run marginal costs. Estimates of average costs can be derived from many sources, including those listed in Table 9.4, which provides a set of estimated shadow prices for various types of alternate programs.

3. Conclusion

In summary, valuing the impacts of the program can be done by using either dollar-denominated outcome measures or shadow prices. In either event, the key issue is reflected in the following guideline:

Guideline 34. Valuing the impacts of programs requires that the data sources, data collection procedures, and estimation techniques produce accurate estimates.

C. Include Intangible Benefits or Costs

We have noted that many program effects often cannot be valued monetarily, but they can be incorporated by the analyst into the benefit–cost framework by

Table 9.4. Estimated Average Monthly Costs for Training, School, Residential, and Income Support Programs

Program	Units	Average cost per month[a]	Data source
Sheltered workshop	Per client	550	U.S. Department of Labor (1977)
Secondary-level classroom education	Per handicapped student	580	Kakalik et al. (1981), Table 2.5
Other school programs	Per handicapped student	320	Kakalik et al. (1981), Dearman & Plisko (1982)
Work–study program	Per trainee	130	Kakalik et al. (1981), Table 2.5
Residential institution or center	Per resident	2,300	Hauber et al. (1984), Table 21
Group home	Per resident	1,400	Hauber et al. (1984), Table 21
Foster care	Per resident	550	Hauber et al. (1984), Table 21
Semi-independent living program	Per resident	930	Hauber et al. (1984), Table 21
Social Security Disability Insurance	Per recipient	470	Social Security Administration (1986), p. 4
Supplemental Security Income	Per recipient	260	Social Security Administration (1985), Table 176
Medicare (disabled persons)	Per enrollee	170	Social Security Administration (1985), Table 137
	Per claimant	290	
Medicaid (disabled persons)	Per recipient	340	Social Security Administration (1985), Tables 153 and 155
Food stamps	Per recipient	43	Social Security Administration (1985), Table 197
Aid to Families with Dependent Children (AFDC)	Per recipient	110	Social Security Administration (1985), Table 195
	Per family	310	

[a]Estimates are expressed in 1985 dollars, with the exception of those for Social Security Disability Insurance and Supplemental Security Income (December 1984), Medicare (FY1982), Medicaid and Food Stamps (FY1984), and AFDC (1983). All numbers are general averages (rounded to two significant digits) for the U.S. as a whole. Costs in specific localities may vary considerably from those presented here. In addition, cost may vary for specific types of services. For example, the costs of many types of post-high-school programs have been aggregated in the category of other school programs. These include college costs of $411 Per month and postsecondary vocational education costs of $182 Per month. As a result, the estimates here should be regarded as indicating the general order of magnitude for cost rather than as precise estimates of actual costs.

including measures of intangible outcomes. For example, self-sufficiency may be assessed by examining the presence of personal benefactors or caretakers, the ability to handle money or travel independently, or moves to less supportive housing environments. Preferences for work may be examined by an individual's level of satisfaction with his or her employment, absenteeism rate, and performance rating. Quality of life will be reflected in part by estimated changes in earnings and any changes in an individual's living situation. It may also be assessed by examining such factors as the number of social events attended, social contacts, and individuals' own reported assessments of the quality of their lives.

The key to including measures of these intangible components is to identify indicators of the relevant concepts. While no explicit monetary value will be assigned to changes in these indicators, they can be seen as evidence of whether the program had the desired effect on intangible outcomes. These changes can also be used to interpret the measured net present value. For example, if measured costs outweighed benefits, then the measured net present value could be thought of as the price of producing the observed change in the indicators of intangible benefits.

This was the approach adopted in an evaluation of the Long-Term Care demonstration project (Thornton & Dunstan, 1986). In that study, it was estimated that the provision of special case-managed community services on top of the existing system of community care increased overall costs to society. However, the provision of these services had several positive effects on the elderly persons who received them. While no explicit value was made of these effects, the evaluation noted that participants had fewer unmet needs, were more confident about their ability to get needed services, and generally expressed greater satisfaction with their lives. Thus, the final conclusion was that total benefits would exceed costs if society valued the increase in well-being by at least the amount by which measured costs exceeded measured benefits. Therefore, we feel that an important guideline relates to including intangible benefits and costs in one's benefit–cost analysis.

Guideline 35. A complete benefit–cost analysis should include both tangible and intangible benefits and costs.

D. Aggregate the Valued Benefits and Costs

The final step in producing a net present value estimate is to determine the appropriate aggregation methods and assumptions. This step entails more than simply summing the estimated value of benefits and costs, because benefits and costs for almost all programs will occur at different points in time. It is not only the magnitude of benefits and costs but also the timing of their occurrence that is relevant in the benefit–cost assessment. To aggregate benefits and costs that occur at different points in time, the analyst must consider three issues—inflation, discounting, and extrapolation.

1. Inflation

To adjust for the effects of inflation, the analyst should denominate all values in dollars from a specified base period. This can be done by valuing benefits and costs based on shadow prices or on cost data that represent market values in the specified base period. It can also be done by adjusting dollar-denominated impact estimates using a price index. One such index is the implicit price deflator for the GNP, which is reported each quarter by the U.S. Council of Economic Advisors. Using a single index, such as the implicit price deflator, has expositional and computational advantages. In addition, a broad-based index such as the GNP implicit price deflator can more accurately reflect price changes in the wide range of commodities that are likely to be affected by social programs.

2. Discounting

All benefits and costs must be calculated in equivalent values by discounting those that occur in later years by a factor that reflects the return that these resources could have earned in the interim between the base period and the time of occurrence. This adjustment is needed because a benefit or cost achieved in a current year is worth more than an identical one achieved later. For example, a monetary benefit that is earned this year could be reinvested and would earn a rate of return over time; consequently, the sum of this benefit and the return on investment would be greater than an equivalent initial benefit achieved later.

Streams of values that have been discounted are termed *present values*. To estimate the present value (*PV*) of, for example, a $1,000 benefit that is expected to occur 10 years from now, the calculation would be as follows, assuming that the rate of return that could have been earned (rate of inflation) on an investment during those 10 years is 5% per year:

$$PV(1+.05)^{10} = 1,000$$

Dividing both sides of the equation by $(1+.05)^{10}$ yields a present value of $614. In other words, a $1,000 benefit earned 10 years from now is worth $614 today, given a discount rate of 5%. By using present values, benefits and costs earned in different years are converted into comparable base period dollars. In addition, values are typically discounted to the same base period used in the inflation adjustments—usually the period during which participants were in the program (and during which most costs were incurred).

The appropriate discount rate to be applied in evaluations of social programs is always somewhat in dispute. While the choice of a discount rate is very important for the evaluation and is well established theoretically, there has never been a completely satisfactory way to estimate discount rates. The fact is that the choice of discount rate typically is made arbitrarily. Most studies of social programs have used rates between 3% and 10%. The 10% rate is mandated by the U.S. Office of Management and Budget (1972) for evaluating government investments. One possible procedure is to assume a middle value, 5%, and then test the sensitivity of the results to this assumption by using 3% and 10% discount rates.

A lower discount rate will increase the present value of benefits or costs that accrue later in time; a higher discount rate will produce a lower present value. Consequently, because the majority of the costs of social programs are incurred during the in-program period (typically the base period to which other benefits and costs are discounted), and because benefits typically accrue later in time, the social net present value will usually change in the opposite direction from the change in discount rate. That is, if a higher discount rate is used, the present value of future benefits will fall, and because costs that accrue near the base period will not be affected very much by the discounting procedure, estimated net present value will decline.

3. Extrapolation

Often the benefits of a program are expected to continue after the observation period during which the impacts of the program are measured. This is particularly true of educational programs, which may produce changes throughout an individual's lifetime. To account for these impacts, the analyst must decide which benefits (or costs) will persist over time, how long they will continue, and at what rate they will persist. For example, if a program is expected to increase a special education student's postprogram output over what it would have been in the comparison situation, the analyst must decide how long this difference in postprogram output will persist. The analyst must also consider whether the magnitude of the initially observed effect will remain at the same level, decline over time, or increase. Thus, in aggregating the valued benefits and costs, don't overlook our 36th guideline:

> *Guideline 36.* Aggregating benefits and costs that accrue at different points in time must include corrections related to inflation, discounting, and extrapolation.

To ignore unobserved impacts is clearly inappropriate; however, the lack of direct observations makes it difficult to estimate the magnitude and value of impacts with confidence. Due to the great amount of uncertainty surrounding each of these questions, we recommend that extrapolations be made on the basis of fairly straightforward methods and relatively simple assumptions that draw on available evidence about long-term effects and that can easily be tested to assess the effect of alternative extrapolation assumptions.

IV. Present and Interpret the Results

Given that much of the value of a benefit–cost analysis stems from the process of organizing and aggregating the data, the presentation of any findings must capture as much of this process as possible. To do so requires that the analyst report more than just the final net present value. It is also necessary to indicate the level of certainty that can be attributed to the analysis and to indicate those benefits and costs that have been omitted from the quantitative estimates. Specifically, the final presentation should include at least three components: (1) a benchmark net present value estimate; (2) inclusion of nonvalued impacts; and (3) a set of alternate estimates based on sensitivity tests.

A. Net Present Value Estimate

The basis for this discussion is a compiled Table 9.2 such as you see in Table 9.7. Presenting the estimated values of benefits and costs in this format provides readers with all the key information in one place. This summary table gives both a quick overview of the results and much of the detailed information necessary for assessing the degree to which individual benefits and costs contribute to those overall results. It also summarizes the general impact of the program on the distribution of resources between participants and the rest of society. We pursue this in more detail in Section V.

B. Nonvalued Impacts

The discussion surrounding the benefit–cost matrix table must also reference any nonvalued impacts that might influence the overall conclusions. Self-esteem, quality of life, health status, and the provision of equal opportunity can be important outcomes of human service programs. While they are generally difficult to value in dollar terms, they cannot be ignored in making the final benefit–cost assessments. These impacts can be included in many ways. One way is to include them as additional rows in the summary table and to record pluses or minuses (depending on whether the nonmeasured item is expected to be a benefit or cost) where the dollar estimates would have gone. This is the approach used by Long et al. (1981) and Kemper et al. (1981).

Another approach is to include in the summary table a brief paragraph outlining the available evidence on these impacts. This was the approach used by Thornton and Dunstan (1986) in the Long-Term Care demonstration project. Their summary table included two panels. One presented the estimated dollar values for those impacts for which dollar estimates could be derived (for example, changes in the costs for hospitals, community services, and nursing homes). The other panel described the estimated impacts on death rates, community residence, the sample members' reports of unmet needs, their satisfaction with life, and the reports of informal care givers (typically, the spouses and children of the elderly sample members) about their satisfaction with service arrangements. The final conclusions were then based on comparisons of the measured net costs and the intangible benefits.

C. Alternate Estimates Based on Sensitivity Tests

The discussion of the summary table must also examine how the overall results would be altered by changes in the assumptions and estimates used to derive them. Such sensitivity tests provide a way to understand the level of confidence that can be placed in the findings and indicate the specific aspects of the program and its evaluation that are most important for producing the overall benefit–cost conclusion.

Some of the types of assumptions or estimates that the analyst may want to test have been mentioned previously. These and others may include the specific shadow prices used, extrapolation factors (for example, assumptions about how long the impacts will persist in the future and at what rates the magnitude of effects will decline or increase over time), and discount rates. They may also include examinations of those estimated effects that appear to have the greatest impact on the final conclusion or those which appear to be particularly uncertain. For example, if postprogram output is expected to be a major benefit, the analyst may want to test the sensitivity of the findings to different assumptions about indirect labor market effects or to alternative methods for estimating the program effect on this output.

The sensitivity tests should be presented in a second summary table. This table would include the initial estimates based on the assumptions that the analyst feels are most accurate and the alternative estimates of the net present value derived in the sensitivity tests. An example of alternate estimates from the STETS analysis, which is discussed in the next section, is presented in Table 9.5.

The analyst may learn that some underlying estimates and assumptions have relatively little impact on whether the estimated net present value is positive or negative and, thus, have minor implications for the overall benefit–cost assess-

Table 9.5. **Alternative Estimates of Net Present Value per Participant (1982 dollars)**[a]

Valuation assumptions	Estimated net present value (in dollars)		
	Social	Participant	Rest-of-society
Basic assumptions	−1,038	2,111	−3,149
Extrapolation beyond 22-month obser-vation period			
7 months; no decay of impacts	4	2,333	−2,329
30 months; no decay of impacts	3,221	3,018	203
10 years; 14% annual decay	5,237	3,447	1,791
Value of Phase 1 and Phase 2 output			
Value = revenue	−3,280	2,111	−5,391
Value = alternative supplier's price	−22	2,111	−2,133
Operational cost equals the cost ob-served for the entire demonstra-tion	−3,552	1,555	−5,107
Discount rate			
3%	−1,025	2,131	−3,157
10%	−1,068	2,060	−3,128
Indirect labor market effects	−2,677	2,111	−4,788

[a]Adapted from Kerachsky *et al.* (1985). See Tables 9.6 and 9.7 for additional details.

ment. Other assumptions may be found to affect the final net present value estimate substantially. Such a finding would suggest that the overall estimates should be interpreted carefully in light of their sensitivity to changes in these assumptions. It may also suggest that, if possible, the analyst should investigate in greater depth the phenomenon underlying any assumption or estimate to which the results are so sensitive.

D. Conclusion

In summary, the final benefit–cost analysis should include those factors listed in our final guideline of the chapter:

Guideline 37. In the final presentation, the findings of the benefit–cost analysis should include (1) a benchmark net present value estimate, (2) incorporation of intangi-ble effects, and (3) a set of alternate estimates based on sensitivity tests.

V. A Benefit–Cost Analysis Example

Previous sections of the chapter have presented our proposed benefit–cost analysis model and its components, using a hypothetical exampe from special education. We now would like to apply that model to actual data from the STETS demonstration project (Kerachsky *et al.*, 1985), which we have used repeatedly in the previous chapters on process and impact analyses. The concepts and definitions defined previously in the chapter are used to elaborate on critical points of the analysis.

A. The Accounting Framework

The accounting framework for the STETS benefit–cost analysis is presented in Table 9.6, which uses the same benefit–cost matrix format as that presented in Table 9.2 for the special education example. Table 9.6 lists the major impact components of STETS (regardless of whether they can be valued) and suggests whether a component is, on the average, a benefit $(+)$, a cost $(-)$, or neither (0), from each of the three perspectives listed—social, participant, and rest of society.

Before proceeding with the separate cost and benefit components, it is important to review how the impact estimates summarized in Chapter 8 (see Table 8.1) are used in the benefit–cost analysis. These estimates indicate the effects of STETS on experimental group members at 6, 15, and 22 months after randomization. These "point-in-time" estimates are adequate measures of the impacts of STETS, but they are inadequate for the benefit–cost analysis, which requires information on the impacts of the program for the entire 22 months. In order to compare benefits and costs, one needs to estimate the cumulative change in earnings, program use, transfer receipt, and other activities. In the absence of continuous data on these activities, it was necessary to derive cumulative measures by interpolating between the point-in-time estimates. The method used was to interpolate linearly between the point estimates. Although any interpolation method involves some arbitrariness, we felt that the estimates of cumulative effects based on linear interpolations provided an accurate indication of the time magnitude of program impacts.

B. Program Costs

The accounting framework disaggregated costs into three components: the operating costs of the projects, compensation paid to participants while they were in Phase 1 or Phase 2 activities (see Table 3.2), and central administrative costs. These cost figures are presented in Table 9.7 (Section I). The operating and central administration costs were paid by nonparticipants (rest of society). Be-

Table 9.6. **Expected Benefits and Costs of STETS by Analytical Perspective**[a]

	Analytical perspective		
Impacts	Social	Participant	Rest-of-society
I. Program costs			
Project operations	−	0	−
Payments to participants	0	+	−
Central administration	−	0	−
II. Output produced by participants			
Phase I and Phase 2 output	+	0	+
Output forgone while in STETS	−	−	0
Increased out-of-program output	+	+	0
III. Other programs			
Reduced use of:			
Sheltered workshops	+	0	+
Work–activity centers	+	0	+
School	+	0	+
Job-training programs	+	0	+
Case-management services	+	0	+
Counseling services	+	0	+
Social/recreational services	+	0	+
Transportation services	+	0	+
IV. Residential situation			
Reduced use of:			
Institutions	+	0	+
Group homes	+	0	+
Foster homes	+	0	+
Semi-independent residential programs	+	0	+
V. Transfer payments and taxes			
Reduced SSI/SSDI	0	−	+
Reduced other welfare	0	−	+
Reduced Medicaid/Medicare	0	−	+
Increased taxes	0	−	+
VI. Transfer administration			
Reduced use of SSI/SSDI	+	0	+
Reduced use of other welfare	+	0	+
Reduced use of Medicaid/Medicare	+	0	+
VII. Intangibles			
Preferences for work	+	+	+
Increased self-sufficiency	+	+	+
Increased variation in participant income	−	−	−
Forgone nonmarket activity	−	−	0
Increased independent living	+	+	+

Note. The individual components are characterized from the three perspectives as being a net benefit (+), a net cost
    (−), or neither (0).
[a]Adapted with permission from Kerachsky *et al.* (1985).

Table 9.7. **Estimated Benefits and Costs of STETS per Participant during the
Observation Period (1982 dollars)**[a]

Impacts	Analytical perspective		
	Social	Participant	Rest-of-society
I. Program costs			
Project operations	−$6,050	$0	−$6,050
Payments to participants	0	3,094	−3,094
Central administration	−182	0	−182
II. Output produced by participants			
Phase 1 and Phase 2 output	3,434	0	3,434
Forgone output while in STETS	−425	−425	0
Increased out-of-program output	268	268	0
III. Other programs			
Reduced use of:			
Sheltered workshops	767	0	767
Secondary vocational school	428	0	428
Other school	112	0	112
Job-training programs	434	0	434
IV. Residential programs			
Reduced use of:			
Institutions	174	0	174
Group homes	72	0	72
Foster homes	7	0	7
Semi-independent residential programs	−114	0	−114
V. Transfer payments and taxes			
Reduced SSI/SSDI	0	−264	264
Reduced other welfare	0	−82	82
Reduced Medicaid/Medicare	0	−232	232
Increased taxes	0	−249	249
VI. Transfer administration			
Reduced use of SSI/SSDI	16	0	16
Reduced use of other welfare	8	0	8
Reduced use of Medicaid/Medicare	12	0	12
VII. Intangibles			
Preferences for work	+	+	+
Increased self-sufficiency	+	+	+
Increased variation in participant income	−	−	−
Forgone nonmarket activity	−	−	−
Increased independent living	+	+	+
Net present value (benefits less costs)	−$1,038	$2,111	−$3,149

Note. Benefits and costs are discounted to the time of enrollment using a 5% real annual discount rate.
[a]Adapted with permission from Kerachsky *et al.* (1985).

cause these costs represent the value of the resources used to operate STETS, they also represent social costs. Participant compensation was treated as a transfer from nonparticipants to participants because it represented a shift in resources from one group to another. It included the wages and fringe benefits paid by the projects and employers to participants in Phase 1 and 2 activities.

C. Output Produced by Participants

The analysis of STETS-induced effects on participant-output distinguished between goods and services produced by participants in Phase 1 and 2 and those produced by them outside of STETS. These two types of output have different distributional consequences and necessitate using different estimation techniques.

a. Phase 1 and 2 Output. The value of the in-program output of participants was estimated on the basis of a series of work-activity case studies for 33 randomly selected experimentals. For each person, the net value of their output during a 2-week reference period was estimated. This estimate was based on the wages and fringe benefits that would have been paid by an employer to other workers to produce the output that was produced by the participant. These "net-value-added" estimates indicated that participants produced output worth an average of $293 per month of active participation during Phase 1 and $503 per month during Phase 2. Since the average length of active participation was 5.5 months in Phase 1 and 3.8 months in Phase 2, the total value of output, discounted to the time of enrollment, was $3,434 per participant.

b. Value of Out-of-Program Output. STETS affected the out-of-program output of participants in two ways: as participants, they forfeited alternative employment opportunities; and as graduates, many were able to work more than would have been the case in the absence of STETS. These changes in output entered the benefit–cost analysis from the perspectives of society and participants. The estimates involved a number of assumptions about indirect labor market effects as described previously in the chapter. The calculations indicated that during the 15 months after randomization, participants forwent non-STETS jobs in which they would have produced output worth $437 per participant. During the 7 months between months 15 and 22, participants produced increased non-STETS output worth $290 per participant. When discounted to the time of enrollment, these estimates implied $425 of output forgone per participant in the first 15 months and a subsequent increase in output worth $268 per participant.

D. Other Benefits

While the primary objective of STETS was to increase employment and earnings, the intervention also generated other important impacts. These are summarized in Table 9.7 (Sections III, IV, V, and VI) within the context of (1) changes in the use of programs other than STETS, (2) changes in the use of residential programs, and (3) changes in government transfers and taxes. In general, these impacts were estimated by multiplying the estimated impact on months by use of a shadow price (such as shown in Table 9.4). For example, changes in the use of sheltered workshops by participants were valued by multiplying the estimated change in the number of months that participants used sheltered workshops by the average monthly cost of operating sheltered workshops. All costs and benefits reflected 1982 dollars.

E. Overall Assessment of Costs and Benefits

The analysis first aggregated the measured costs and benefits which are summarized in Table 9.7. The analysis then considered the intangible (unmeasured) benefits and costs pertaining to changes in employment, social integration, and independence. Finally, the analysis included additional sensitivity tests such as those reflected in Table 9.5.

a. *Measured Costs and Benefits.*   The estimates presented in Table 9.7 suggest that STETS created a net cost to society during the 22-month observation period. The measured social costs totaled $6,232 per participant, while measured social benefits (increased output by participants and the reduced use of other training, service, residential, and transfer programs) totaled only $5,193 per participant. Thus about 83% of the initial investment was offset during the 22-month observation period. Participants clearly benefited from their participation, receiving in-program compensation that more than offset their tax payments and their reduced use of transfers. Nonparticipant taxpayers incurred the costs both for operating STETS and for participant compensation. They received substantial benefits (primarily from the increased output produced by participants in STETS and the reductions in their use of sheltered workshops, other job-training programs, and transfer programs), but these benefits offset only two-thirds of the costs incurred by nonparticipants. However, trends observed for the impacts on earnings and the use of sheltered workshops suggested to the analysts that benefits would persist and were likely to outweigh costs in the long run. For example, if the earnings and reduced alternative program benefits continued for as little as 7 months beyond the 22-month point, social benefits would exceed social costs.

b. *Intangible Effects.*   The STETS program was also intended to enhance the economic and social self-sufficiency of participants (see Tables 9.6 and 9.7, Section VII) The measured impacts indicated that STETS did affect the activities and opportunities that were expected to generate intangible benefits. The increased income and the increased job-holding in the regular labor market provided participants with benefits that went beyond the measured increases in output. Indeed, increased social and employment opportunities were available to the participants. However, the analysis found limited evidence of changes in such intangibles as self-sufficiency and independence. In part, such limited evidence reflects the inadequacies of the measures and the difficulty in measuring these concepts. It may also mean that self-sufficiency responds slowly to changes in opportunities, particularly for mentally retarded young adults. These persons may feel that they must maintain their jobs and their increased social interactions for a considerable time before they alter their behavior in terms of residential situation, benefactors, and financial independence. Finally, there were no measures of any overall increases in satisfaction, other than the fact that many participants appeared to remain voluntarily in their jobs.

Changes in earnings can also create intangible costs. For example, the stress of employment may create health problems. Moreover, persons who lost their eligibility for transfers such as SSI or SSDI potentially faced more uncertainty about future income than they would have had they remained in those programs. However, it was presumed that participants valued the increased earnings and interactions more than the costs of the intangibles, because they made the choice to enter STETS, and many continued in their jobs after leaving STETS.

F. Conclusion

We hope that this brief benefit–cost analysis example demonstrates how the proposed benefit–cost model can be applied to various social programs. However, we again wish to stress that the benefits and costs are based on several assumptions and estimates that, while plausible, introduce unavoidable uncertainty into the benefit–cost assessment. Thus, one should be aware of the following caution regarding the need for conducting sensitivity tests. The general procedure is to change one underlying assumption while keeping all others the same.

Caution: Because of the uncertainty in benefit–cost analyses due to the numerous assumptions and estimates employed, it is essential to conduct sensitivity tests to examine alternative benefit and cost calculations that incorporate different sets of assumptions.

VI. Summary

In this chapter we discussed our approach to benefit–cost analysis by outlining the critical components of a benefit–cost model and then demonstrating how the model can be used to analyze human service programs. As we saw, benefit–cost analysis is both a structured comparison and a process for weighing a program's benefits or costs. The process starts with a comprehensive accounting framework that includes all benefits and costs, regardless of whether they can be measured or valued. The analysis then proceeds to value as many of those benefits and costs as it can. The number of items that can be valued will depend on the resources available to the analyst, the estimation methods employed, and the nature of the program itself. The analyst then attempts to assess the remaining items that were unmeasured. Some of these unmeasured items can be assessed by examining measures that are closely related to the item of interest. For example, while many programs are expected to improve the quality of life for participants, it has been extremely difficult to include dollar measures of this effect in benefit–cost analyses. Nevertheless, it is often possible to include measures that indicate the general nature of effects on the quality of life, measures based on the stated attitudes of participants or on observations of things such as their range of activities, health status, or unmet needs. When such indicators are unavailable, the analyst can only list the unmeasured benefits and costs and make conjectures about their potential implications for the findings based on measured items.

When this approach is used, decisions can be made using all the available information. Further, the inclusion of all effects, regardless of whether they are explicitly measured, ensures that the decisions will be made with a comprehensive view of the program; important, but intangible, benefits will not be excluded when decisions are made. Even when there are important items left unmeasured, the benefit–cost analysis simplifies the decision by summarizing the information about measured items so that they can be more conveniently compared with the information about unmeasured items.

In this approach, the analyst considers whether the conclusions would be changed if the unmeasured items were explicitly valued and included in the analysis. If the measured benefits outweigh the measured costs, then overall net present value will be positive, unless there are offsetting unmeasured net costs. If measured benefits fall short of measured costs, then the resulting net measured cost can be viewed as the price that must be paid to obtain any unmeasured net benefits.

Another method of dealing with the uncertainty inherent in the analysis is to make several alternative estimates of net present value. This set of estimates includes a benchmark estimate, which incorporates the assumptions and estimates with which the analyst feels most comfortable, and several other estimates based on sensitivity tests, each of which illustrates the effects of changing one or more of the assumptions or approximations used in the benchmark calculation. The process of producing these alternative estimates, identifying the various impacts and outcomes, integrating measures of them, and noting the general patterns that emerge from attempts to assign relative values is quite useful. By basing the findings of a benefit–cost analysis on this range rather than on a single set of assumptions and estimates, it is possible to identify both those aspects of the program and its evaluation that are most crucial and those about which the greatest uncertainty exists. Such an understanding of the program, its performance, and its assessment is crucial to rendering valid cross-program comparisons and decisions about the trade-offs involved in alternative funding.

While benefit–cost analysis seems quite useful for summarizing information and addressing both efficiency and equity issues, it has been adopted slowly by persons studying human service programs. In part, this reflects a sense that the emphasis benefit–cost analysis has traditionally placed on efficiency rather than equity is inappropriate for most human service programs. It also reflects a concern that many effects of intervention or treatment will be missed in a benefit–cost analysis that focuses only on items that can be valued easily in dollars. Finally, it reflects the complexity of many of the techniques and the dearth of useful paradigms for conducting a benefit–cost analysis of human service programs.

The lack of paradigms is a particularly important problem. Such models are important to help persons interested in the field of human services conduct benefit–cost analyses and interpret the findings of ones they read. Unlike formal statistical analysis, there is no uniformly accepted set of rules for conducting a benefit–cost analysis. There are general guidelines, but each analyst generally makes his or her own decisions about how to assess whether omitted effects and costs will affect the measured results. This lack of precise rules, along with differences in evaluation budgets, the expertise of the analysts, and the evaluation goals, leads to substantial variation in the analysis methods used. This variation, in turn, makes it difficult to compare studies and programs, since it is difficult to know if differences in estimated benefits and costs are due to real program differences or merely to differences in evaluation technique. Such confusion defeats one of the purposes of benefit–cost analysis, which is to make cross-program comparisons easier by measuring all effects in the common denominator of dollars. We return to this issue in the next chapter.

In the end, benefit–cost analysis must be seen as a process for organizing information. It is not an inflexible rule that can be used to make decisions. Rather, it helps an analyst to sort through a wide variety of data and to aggregate them so that decisions can be made more easily. In particular, it provides a convenient summary measure for those impacts that can be measured and valued in dollars. It also provides a framework for assessing the potential importance of impacts that cannot be valued, and the uncertainty surrounding the various impacts that can be valued. Policymakers must still make the decision, based on the available evidence and their value judgment. However, it is hoped that by making the systematic comparisons of a benefit–cost analysis, better decisions can be made more easily.

VII. *Additional Readings*

Amemiga, T. (1985). *Advanced econometrics.* Cambridge, MA: Harvard University Press.

Collingnon, F. C., Dodson, R., & Root, G. (1977). *Benefit–cost analysis of vocational rehabilitation services provided by the California Department of Rehabilitation.* Berkeley: Berkeley Planning Associates.

Gramlich, E. M. (1981). *Benefit–cost analysis of government programs.* Englewood Cliffs, NJ: Prentice-Hall.

Lansing, J. B., & Morgan, J. N. (1971). *Economic survey methods.* Ann Arbor, MI: Institute for Social Research, The University of Michigan.

Levin, H. M. (1975). Cost-effectiveness analysis in evaluation research. In M. Guttentag & E. L. Struening (Eds.), *Handbook of evaluation research, Vol. 2* (pp. 89–124). Beverly Hills, CA: Sage.

Noble, J. H. (1977). The limits of cost–benefit analysis as a guide to priority-setting in rehabilitation. *Evaluation Quarterly, 1,* 347–480.

Rossi, P. H., & Freeman, H. E. (1982). *Evaluation: A systematic approach* (2nd ed.). Beverly Hills, CA: Sage.

Sorensen, J. E., & Grove, H. D. (1977). Cost-outcome and cost-effectiveness analysis: Emerging nonprofit performance evaluation techniques. *The Accounting Review, 52*(3), 658–675.

Thompson, M. D. (1980). *Benefit–cost analysis for program evaluation.* Beverly Hills, CA: Sage.

Warner, K. E., & Luce, B. R. (1982). *Cost–benefit and cost-effectiveness analysis in health care.* Ann Arbor, MI: Health Administration Press.

10

Back-of-the-Envelope Benefit–Cost Analysis

I. Overview

We stressed throughout Chapter 9 that benefit–cost analysis should be perceived as a process for organizing information rather than as an inflexible rule for making decisions. In particular, benefit–cost analysis provides a convenient summary measure for those impacts that can be measured and valued in dollars and a framework for assessing the potential importance of impacts that cannot be valued in dollars.

The examples we have presented require considerable resources to collect data, estimate impacts, and conduct the analysis. But, as we promised in the overview to Section IV, we want to suggest to the evaluation producer a technique that can be used with limited resources and that will be useful to the program administrator. The technique has two steps. First, get the comparison situation from the setup (Chapter 2) and the cost estimates for the program (Chapter 6), and then complete an accounting framework. Second, do implied benefits by working backwards to figure out how big the impacts would need to be for the net benefits to exceed costs.

Although we ask you to be precise in the analysis, we realize there will be considerable uncertainty in your estimates because of possible lack of impact statements. However, the critical thinking involved in the process and what the analysis might well suggest about your program will be well worth the effort. But first, we have a caution:

Caution. Back-of-the-envelope benefit–cost analysis results in considerable uncertainty and lack of precision. Be very cautious in your extrapolations.

II. Develop an Accounting Framework

Developing an accounting framework focuses on the conceptual level. It stresses the value of taking time to develop an accounting framework for your program and of paying attention to the program's various perspectives, costs, and benefits. Survey 10.1 can be used to develop your accounting framework. Most of the information that you need to include under benefits, costs, and analytical perspective can be obtained from previous tables for program outcomes (Survey 7.1), costs (Table 6.1), and anticipated impacts (Table 8.1). We discussed perspectives in Chapters 8 and 9, but we need to make a few additional comments to help you think through how the anticipated impacts of the program are expected to be perceived (that is, as benefits or costs) by the different groups.

The benefit–cost accounting matrix you are asked to complete in Survey 10.1 imposes a logical rigor on the analysis and serves as a guide for interpreting the results. The framework specifies a consistent method for valuing the diverse set of effects by focusing on the net resource gain or loss induced by your program as it is implemented. As we discussed in Chapter 9, the approach essentially entails estimating the change in resources available because of your program and then valuing those resources at their market cost.

While the above procedure assigns a value to program effects, that value will be viewed differently by different groups. For example, if participants lose their SSI benefits as a result of getting a job because of your program, they will view the loss as a cost, while taxpayers ("rest of society") will view it as a financial gain. The survey captures these differences through three analytical perspectives: society as a whole, program participants, and nonparticipants, whom we refer to as "the rest of society." This nonparticipant group includes everyone in society who is not given the opportunity to participate in your program. It therefore encompasses much more than the control group, which comprises a very small part of the nonparticipant group. We prefer this term to the more common "taxpayer group," because the participant group frequently pays taxes and because not all of the effects on nonparticipants occur through the tax system. Those three analytical perspectives additionally result in an analytically useful feature: the sum of the net present values calculated from the participant and nonparticipant perspectives equals the net present value for the social perspective. This "adding-up" property is valid because participants and nonparticipants constitute mutually exclusive groups that when combined include all members of society. Therefore, transfers of income between these two groups cancel each other out in the social perspective, because the benefit to one group is assumed to be equal to the cost to the other.

This adding-up property is clearly seen in Figure 1.1 (Chapter 1). Note that each of the three programs has defined target populations that constitute mutually exclusive groups. When combined, these groups include (or add up to) everyone

Survey 10.1. **Benefit–Cost Accounting Framework Table Shell**

| Impacts | Analytical perspective[a] | | |
	Social	Participant	Rest-of-society
Benefits[b]			
1. Increased output			
a.			
b.			
c.			
2. Reduced use of alternative programs			
a.			
b.			
c.			
3. Reduced use of transfer programs			
a.			
b.			
c			
4. Other benefits			
a.			
b.			
c.			
Costs[c]			
1. Program costs			
a.			
b			
2. Forgone market output			
3. Increased use of complementary programs			

[a]These are only suggested perspectives. Use others if more appropriate to your program.
[b]These benefit categories are only suggested. Develop other categories if more appropriate to your program. See Tables 9.2 and 9.6.
[c]These cost categories can be modified to meet the needs of your analysis. See Tables 9.2 and 9.6.

(all members of society). If one changes the use of either of the programs, it will affect nonparticipants, but the adding-up property will still apply. Similarly, if we introduce Program A and evaluate it versus the situation without A, the sum of the net present values calculated from the participant and nonparticipant perspectives will still equal the net present value for the social perspective. We encourage you to use Figure 1.1 in this way to show or to better understand the different analytical perspectives.

With your increased understanding of the three perspectives, you now need to complete Survey 10.1 by evaluating whether a component is, on the average, a benefit $(+)$, a cost $(-)$, or neither (0), from each of the three perspectives. Once the evaluations are made, spend some time analyzing what the pluses and minuses mean (we suggest you ignore the zeros, since they represent no difference in impacts). The number of pluses and minuses might not be as important as their magnitude. However, things you might want to focus on in your analysis include:

1. Who gains and loses the most in reference to the three perspectives?
2. What are the incentives and disincentives to someone entering your program?
3. How much of your scoring and analysis is based on certainty, and how much is based on uncertainty?

In summary, although we realize that this first step is not sophisticated, we do feel that developing the analytical framework for your program will result in a number of benefits. First, it forces you to look at the various perspectives and evaluate what each gains and loses from the program. Second, it helps you to focus on incentives for entering or not entering your program. For example, glance down the participant column and see where the pluses and minuses are. Numerous minuses may account for high attrition rates, low placement rates, or high program refusal rates. And third, we feel that the process of putting the chart together will result in critical thinking about the program's costs and benefits and how to increase either the program's efficiency or equitableness.

III. Implied Benefit Analysis

The second step of our suggested technique involves doing implied benefits by working backwards to figure out how big the impacts would need to be for the net benefits to exceed costs. The suggested implied benefits strategy can be used if you have *assumed* that benefits equal costs and you know what the costs are. The question then becomes, how big are the implied benefits? The procedure involved in using this strategy includes the following: (1) decide on the time horizon (for example, a 5-year payback); (2) decide on the discount rate (assume 5%); (3) decide what benefits to include (for example, earnings plus change in other programs); (4) obtain your cost estimates, given what you know about average costs and shadow prices from Chapter 6; (5) do the calculations; and (6) compare the implied benefits to what is known about the population to get some idea if it is feasible to get benefits of that magnitude.

For example, assume you administer a job placement program that costs

$1000 per person placed. The program claims that it can pay for itself in a social sense. Furthermore, you need to recoup the benefit in 5 years. Therefore, the participant needs to earn back close to $300 per year (with discounting) before the benefits will exceed the costs. Then take the $300 and compare with T_B (what the participants were earning at baseline). Could you achieve the $300, and does that seem plausible? If the participant's income exceeds the necessary amount, then it shows that the program has some potential, which is very important during the program's feasibility stage. But since the analysis was done on the back of an envelope, all you can say is that it seems to have potential.

One challenge to using the implied benefits approach is that benefits and costs may occur over a long time period. As pointed out in Chapter 9, discounting is a procedure for adjusting the values of benefits or costs that occur over time so that all values are expressed in their equivalent value in a specified base period. This procedure is necessary so that programs with different time patterns of effects can be compared directly.

When making a back-of-the-envelope benefit–cost analysis of a program that produces a stream of net benefits into the future, it is useful to have a method that will approximately account for the process of discounting. Table 10.1 provides a set of factors that can be used in this way. They can help you determine the required annual net benefits that will be necessary to have total benefits outweigh costs.

Table 10.1. **Discount Factors for Back-of-the-Envelope Benefit–Cost Calculations**

Annual discount rate	Time horizon (years)					
	5	10	15	20	25	30
0	5.00	10.00	15.00	20.00	25.00	30.00
1	4.88	9.52	13.93	18.13	23.12	25.92
2	4.76	9.06	12.96	16.48	19.67	22.56
3	4.64	8.64	12.08	15.04	17.59	19.78
4	4.53	8.24	11.28	13.77	15.80	17.47
5	4.42	7.87	10.55	12.64	14.27	15.54
6	4.32	7.52	9.89	11.65	12.95	13.91
7	4.22	7.19	9.29	10.76	11.80	12.54
8	4.12	6.88	8.74	9.98	10.81	11.37
9	4.03	6.59	8.23	9.27	9.94	10.36
10	3.93	6.32	7.77	8.65	9.18	9.50
11	3.85	6.06	7.35	8.08	8.51	8.76
12	3.76	5.82	6.96	7.58	7.92	8.11

Note. The factors indicate the present value of a stream of $1 payments for different lengths of time and discount rates.

The factors in Table 10.1 show, for several combinations of time horizon and discount rate, the discounted value of a stream of benefits that produced $1 per year. Analysts should select a time horizon that is the shorter of two time periods: (1) the length of time that program impacts are expected to last, or (2) the length of time you or other policymakers are willing to wait before requiring that all program costs be recouped. The discount rate is essentially an interest rate indicating the extent to which current and future dollars can be traded (for example, a 5% discount rate indicates that society demands $105 next year in order to give up $100 in the current period). In most cases, a 5-year time horizon and a 5% annual discount rate will provide a good basis for assessing potential benefits and costs.

Once choices have been made about the appropriate discount rate and time horizon, the appropriate back-of-the-envelope discount factor can be read from the table. As an example, consider a net benefit stream that produces $1.00 a year for 5 years. If your discount rate was 5%, then the table indicates that the present value of that stream of dollars would be $4.42. The table also indicates that the value of future benefits falls as the time horizon gets shorter or the discount rate gets higher. In the example, the present value of the stream of $1.00 payments would be only $3.93 instead of $4.42, if the discount rate increased from 5% to 10%.

In applying these factors, the analyst would divide the estimate of program costs by the relevant factor. The analyst would then have to determine the likelihood of the program generating a stream of net benefits equal to that value. If we maintain our assumption of a 5-year time horizon and a 5% discount rate, then we would divide the program costs by $4.42. Thus, a program that cost $4,000 per participant would have to generate net benefits of at least $905 a year for 5 years if it is to be cost beneficial. Longer time horizons or lower discount rates imply that the required annual net benefits would be lower; shorter time horizons and higher discount rates imply the need for greater annual benefits. For example, the same program would only need to generate annual benefits of $280 if there was a 25-year time horizon rather than a 5-year horizon.

Calculations made in this way are necessarily approximations, but they are useful for back-of-the-envelope comparisons of benefits and costs. Typically, net benefits will not occur in equal amounts over time. Many programs incur most of the costs in the first year and then produce net benefits for some time after. Benefit–cost calculations using the factors in Table 10.1 therefore offer only a rough approximation to the actual value. Nevertheless, this approximation can be derived relatively easily and can be used to assess the likelihood that a program will generate benefits that exceed its costs. In many cases, this rough assessment will be sufficient. In other cases, it will indicate that a program has the potential to produce benefits in excess of costs, and that further research is needed in order to determine the extent to which that potential is realized in practice.

We purposefully suggest that other, intangible benefits should probably not be analyzed on the back of an envelope because of the probable lack of impact estimates and the inappropriateness of using a broad brush approach in their estimation. However, their omission from this chapter should not suggest their insignificance. Indeed, we emphasized throughout Chapter 9 that many program effects often cannot be valued monetarily but, with careful planning, can be incorporated into a larger benefit–cost framework by including measures of intangible outcomes. The key to including measures of these intangible components is to identify indicators of the relevant concepts. For example, self-sufficiency may be assessed by the presence of personal caretakers, the ability to handle money or travel independently, or moves to less-supportive housing environments. Preference for work may be examined by an individual's level of satisfaction with employment, absenteeism rate, or performance rating. Quality of life may be reflected in part by changes in earnings, community involvement, and changes in an individual's living situation. If impact estimates of intangibles such as these are available, then they can readily be included in your accounting framework (Survey 10.1).

IV. Summary

This chapter addressed the needs of the evaluation producer or program administrator who wants to attempt a benefit–cost analysis, but who has limited resources, cost estimates, or impact estimates. We outlined a suggested technique that employs a broad brush approach but is suitable for a back-of-the-envelope benefit–cost analysis. We caution, however, that such an analysis results in considerable uncertainty and lack of precision.

The first step in the proposed approach involves developing an accounting framework, which includes evaluating a program's benefits and costs from three analytical perspectives—society, the participant, and the nonparticipant (the rest of society). The accounting framework and subsequent evaluation allow one to think critically about who gains and loses from the program and what are the incentives and disincentives to someone who is eligible for the program. The second step involves doing implied benefits, which means determining how big the impacts would need to be for the net benefits to exceed costs.

In the end, benefit–cost analysis must be seen as a process for organizing data rather than an inflexible rule that can be used to make decisions. We hope this chapter has assisted the program evaluator or administrator to sort through a wide variety of data and to aggregate them so that decisions can be made more easily. Policymakers must still make the decision, based on the available evidence and their value judgment. However, we trust that this chapter on back-of-the-envelope benefit–cost analysis will help you better understand your program and make better decisions about it.

V

Analysis to Action

Decision makers and program analysts often face a mutual Catch-22: despite the increasing number of program analyses, one frequently finds reluctance to implement the results. Although reasons vary, the reluctance is generally attributed to organizational, bureaucratic, political, or credibility factors. But these are not the issues we specifically want to focus on in this last section of the book. Rather, we want you as a producer or consumer to think about some ways to increase your communication skills, for each of us has a primary responsibility to communicate clearly and concisely to our audience(s). In fact, we feel that communication is the most important tool of our trade. Unfortunately, the need for clear and concise communication is often hampered by lack of training, clear thinking, or interest on the part of our readers. The net result frequently is that a lot of time, effort, and other resources go into program evaluations that people don't read. Thus, conducting program analysis along the dimensions outlined in previous chapters is only part of the process; one must also translate the results of those analyses into action. Suggestions for doing so are the focus of this section of the book.

Throughout the book, we have presented guidelines and steps for conducting program analysis. We additionally have stressed that program analysis does not occur in a vacuum; one must constantly be sensitive to the target audience and the need to establish one's credibility by clearly describing the procedures followed, the assumptions made, and the limitations of the analysis. Our observation has been that policy and decision makers become skeptical about the results of program analysis when the analyst overstates or overpromises. We have also observed that analysis results need to be presented clearly, attractively, and honestly.

The section contains two chapters that are based on our experiences with policy and decision makers and the literature dealing with marketing strategies and organizational change. In Chapter 11, entitled ''How to Communicate Your Findings,'' we focus primarily on the need to establish one's credibility, to be sensitive to the audience's needs, and to present the analysis in an attractive, easy-to-read format. Throughout these sections of the chapter, we stress the importance of style, content, and clear communication. Our premise is that reporting the program analysis in a credible, easily (and quickly) understood format is essential for its implementation.

But good communication is only part of the analysis to action; it also requires a concerted effort to translate program analysis results into potential program change. And this is the focus of the last part of Chapter 11, in which we suggest that one should communicate from a marketing perspective and incorporate good marketing principles into one's efforts to implement necessary programmatic changes. The material and techniques presented in this part of the chapter draw heavily from the marketing and organizational change literature. Our premise is that by using marketing techniques, program analysis results are more likely to be acted on, rather than referred to the circular file.

Chapter 12 represents both an epilogue and a challenge. In the chapter, entitled "How Am I Doing?" we present a self-survey you can use to answer that question. In the chapter we also summarize the major themes stressed throughout the book and challenge you to become involved in program evaluation.

Throughout the two chapters, we build on the material presented in previous chapters and suggest that program evaluation is within the grasp of all administrators. In addition, we suggest that one cannot overlook the contexts within which program evaluations occur. For example, Weis (1973), in writing about the political nature of program evaluation, argues that attempts to solve social problems are forged in the political arena, complete with all the vested interests and influences involved in the world of decision making. Similarly, Edwards, Guttentag, and Snapper (1975) state:

> Given the muddiness and complexity of organizational and bureaucratic practices, the separation between researchers and political decision makers and the potent political context of many programs, the core problems of evaluation [are] in the large . . . gap between evaluation data and the decisions of policy makers. (p. 140)

We hope that these two chapters will help bridge that gap and assist decision makers and analysts alike translate program analysis results into action.

11

How to Communicate Your Findings

I. Overview

Most readers have undoubtedly experienced the frustration of either completing a program analysis that collected dust on someone's shelf or reading an analysis report and not knowing what to do with it. This is an all too common scenario in the world of program analysis. Although we offer no panacea, we do feel that how you present the analysis and its results to the target audience can significantly affect its acceptance and potential implementation. To that end, this chapter is divided into four sections that outline various techniques you might use—primarily as an evaluation producer—to communicate your findings. The first section deals with establishing your credibility. The second deals with being sensitive to the needs of the target audiences and the context within which these audiences operate. The third involves presenting your results in an attractive, clearly understandable and readable form, and the fourth involves communicating from a marketing perspective. One of the primary motives for doing program analysis, or reading about it, is to enable administrators to make better decisions. Many of those decisions result in changing programs or policies. Hence, your findings should be communicated within the framework of good marketing principles that will help promote that change.

The chapter addresses the needs of both the evaluation producer and consumer. For the producer, we discuss specific techniques for establishing credibility and presenting results in an attractive, clearly understandable and readible form. But the consumer also needs to use these techniques, along with sensitivity to the needs of the target audience and the use of marketing principles to translate into action the results of others' analyses.

II. Establishing Credibility

One's credibility involves more than a graduate degree. Indeed, it involves logical reasoning, presenting the assumptions made in the analysis, being honest about what actually was done in the analysis, marketing the facts carefully, and pointing out potential pitfalls in the respective analysis. Additionally, not all human service programs have the capability to do process, impact, or benefit–cost analysis. As stated earlier, our suggested approach to program analysis does not ask that programs being evaluated use all three methods or that all aspects of the program be quantified; what we do ask is that analysts be systematic in their collection and analysis of program data, and that their conclusions be based on thoughtful, correct, and honest analysis. It is within this context that we would like to resensitize the reader to the guidelines governing program evaluation discussed in Chapter 3. These rules are listed in Survey 11.1. We refer to them within the context of this chapter as factors influencing one's credibility.

As we have stressed repeatedly, program evaluation is primarily a structured comparison. As a first step in the evaluation, one needs to specify the program or policy being evaluated and the program or option with which it will be compared. This specification should include information on such factors as the persons being served, the treatments being offered, and the environment within which the program or policy operates. These two alternatives—the program and the comparison situation—define the scope and ultimately the results of the evaluation.

In Chapter 3, we discussed those guidelines/credibility factors listed in Survey 11.1. The setup includes the goal of the evaluation, the context within which the evaluation is to be interpreted, and the program model. The major purpose of the setup is to define the structured comparisons, pose the analytical questions, and rationalize the expected outcomes from the services provided. Marshalling the evidence relates to providing evidence to justify or support the contention that the program described produced the desired effects. Once the evidence is marshalled, it is interpreted within the context of the structured comparisons, the program's environment, and any limitations or weaknesses of the evaluation performed.

We now suggest that you refer to Survey 11.1 and evaluate the program analysis in question. Simply answer yes or no to each of the credibility factors listed. We are not proposing a passing or a failing grade; but we do suggest that your credibility is increased proportionally to the number of factors to which you answer yes. The tabled survey can also be used by the evaluation consumer to determine the credibility of an evaluation or analysis read.

III. Being Sensitive to the Target Audience

Program analysts must deal with a number of audiences in their efforts to inform the various constituents of any human service program. Decision makers

Survey 11.1. **Factors Influencing One's Credibility**[a]

Analysis component	Credibility factors	Included in the analysis[b]	
		Yes	No
The setup	1. Problem addressed 2. Persons served 3. Services provided 4. Evaluation context 5. Expected outcomes 6. Justification (rationale)		
Marshalling the evidence	1. Outcome defined 2. Measures selected that reflect initial intent 3. Geographical context of program described 4. Services delivered measured 5. Changes in outcomes measured 6. Data collected on both sides of the structured comparisons		
Interpreting the findings	1. Correspondence between setup and marshalling the evidence 2. Change in outcomes attributed to the intervention 3. Conclusions interpreted within the program's context 4. Issues of uncertainty addressed		

[a]Adapted from Table 3.1 and Survey 3.1.
[b]Place a check beneath either Yes or No.

can include a number of groups, each of which has its own needs, such as those listed below:

- *Federal and state policymakers,* who are interested in efficiency and equity issues and in whether the particular program is accomplishing its goals and at what cost.
- *Program operators,* who are interested in tracking resources, in services provided to participants and the outcomes of these services, and in refining service designs to maximize project effectiveness.
- *Consumer groups and participants,* who are frequently interested in expanding options and in choosing the best program for themselves or their target audience.
- *Social scientists,* who are frequently interested in tracing the net impacts of the program as well as in analyzing the factors influencing project effectiveness.

In addition, numerous authors (cf. Attkisson *et al.,* 1978; Bryson & Cullen, 1985; Hanson, 1979; Hellreigel & Slocum, 1982) have suggested that communicating one's findings can be maximized by focusing on a number of contextual variables that impact the target audience. A number of these are listed in Table 11.1.

As the above information suggests, as evaluation producer or consumer, you need to be aware of the audiences and contextual variables that influence both your communication and the acceptance of the program analysis. Our best advice regarding the audiences and contextual variables is to be sensitive to them and realize that your marketing strategies will probably need to address them. We feel that this aspect of communicating your findings is important enough to suggest the following caution:

Table 11.1. **Contextual Variables Influencing the Target Audience**

Agreement with the analysis effort.

Resources involved including time, money, staff, and expertise.

Use of the data and its impact on resource allocation and organizational structure.

Impact of the evaluation and proposed changes on the organization's ongoing activities.

Attitudes of involvement and commitment from key persons.

Open communcation between evaluator and program personnel.

Built-in review and modification procedures.

Assurance that the analysis is sensitive to the desired outcomes, needs, and concerns of key persons.

Endorsements of analysis plan by influential components of the program's constituency.

> *Caution.* In communicating your findings, do not for-
> get the various groups involved and the contextual vari-
> ables that will influence the target audience.

IV. *Presenting the Results*

The recent management literature with its emphasis on the search for excel-
lence, the skeptical mind, and the one-minute decision maker suggests something about the reporting of analysis results. Analysis and writing are inextrica-
bly entwined such that, "at the bottom, analysis is thinking and writing clearly"
(Murphy, 1980, p. 131). The role that a well-written and -packaged report plays
in translating program analysis results into action cannot be overemphasized.
Decision makers are skeptical and busy; hence, apart from a physically attractive
and easily readable report, you might also keep the following "product criteria"
in mind when writing or reading the final report. Affirmative answers to these
eight questions will greatly increase the product's marketability:

1. Is the report plausible?
2. Is the report coherent?
3. Do the facts of the report correspond to known facts?
4. Is the report adequately documented?
5. Do the analysts take into account alternative interpretations?
6. Are the analysts straightforward about the evaluation's limitations?
7. Did the analysts use sound procedures?
8. Do the conclusions and recommendations follow logically from the information and data presented?

In addition to these questions, both the report writer and reader generally
have questions about the report's style and content. Four frequently asked questions are (1) what should the analysis report contain, (2) what type of writing
style should be employed, (3) how should tables be presented, and (4) what type
of graphics should one employ?

a. Report Contents. Although the report's contents will vary depending
upon the type of analysis, questions asked, and the audience, we feel any report
should contain a number of critical components, such as those listed in Table
11.2. The report should include a rationale statement that relates to the purpose
of the analysis and to the public policy context within which it operates. The next
section should describe the analysis design, including the hypotheses tested and a

Table 11.2. **Critical Components to an Analysis Report**

A. Rationale statement
   1. Public policy context
   2. Program model with goals and objectives
B. Analysis design
   1. Analytic approach
   2. Hyotheses tested
   3. Description of control/comparison groups
C. Process
   1. Persons served and services received
   2. Internal and external organizational factors
   3. Estimates of program cost
D. Estimates of program impacts
E. Public policy considerations
   1. Program potential
   2. Benefits and costs
   3. Generalizability of the findings
   4. Threats to internal validity

description of the control or comparison groups. The process section should contain a description of the persons served, internal and external organizational factors (see Figure 5.2), and estimates of program costs. The fourth section should include an estimate of program impacts. The final section should deal with public policy considerations, including program potential, benefits, costs, generalizability of the data, and any shortcomings of the study.

b. Writing Style. Nothing is more frustrating than trying to read something that is boring, lengthy, or noncommunicative—and that is equally true of an analysis report. Thus, our advice here is to follow a number of suggestions discussed in detail in a delightful little book by Strunk and White (1979) entitled *The Elements of Style.* The authors' 21 suggestions regarding writing style are listed in Table 11.3.

They suggest that one's writing style should be simple and jargon free, clean and uncluttered, well organized, interconnected, responsive, and lively. In addition, Donald McCloskey, in an article on economic writing (McCloskey, 1985), suggests five additional standards for an effective writing style:

- Writing is the analysts' trade: the hard business of writing is to marshall ideas well.
- Writing is thinking. Writing resembles mathematics; if mathematics is a language, so too is language a mathematic—an instrument of thought.

Table 11.3. **Suggestions regarding Writing Style**[a]

---

1. Place yourself in the background and draw the reader's attention to the sense and substance of the material.
2. Write in a way that comes naturally.
3. Work from a suitable design and outline.
4. Write with nouns and verbs.
5. Revise and rewrite.
6. Do not overwrite by using rich, ornate prose.
7. Do not overstate.
8. Avoid the use of qualifiers.
9. Do not affect a breezy manner. Be succinct.
10. Use orthodox spelling.
11. Do not explain too much.
12. Do not construct awkward adverbs by adding -ly to them.
13. Make sure the reader knows who is speaking.
14. Avoid fancy words.
15. Do not use dialect unless your ear is good.
16. Be clear.
17. Do not inject opinion.
18. Use figures of speech (similes, metaphors) sparingly.
19. Do not take shortcuts at the cost of clarity.
20. Avoid foreign languages.
21. Prefer the standard to the offbeat.

---

[a]Adapted with permission from Strunk and White (1979).

- Rules can help, but bad rules hurt; use reasonable rules such as those presented in Strunk and White (1979) and Williams (1981).
- Be thou clear: the rule of clearness is to write not so that the reader can understand but so that he cannot possibly misunderstand.
- The rules are empirical: good style is what good writers do.

c. Use of Tables.   Tables can be very effective in communicating information, provided they are well constructed and understandable. A good example is presented in Table 11.4. Note the clarity and ease of reading the tabled data. In considering the use of tables, the *Publication Manual of the American Psychological Association* (1984) contains a number of helpful suggestions, including those listed in the top portion of Table 11.5.

We feel tables are an effective way to either summarize data or present information in a format that is easy to read, comprehend, and use. We have found that people will use visual aids; we have enjoyed watching persons in libraries or at publishing company exhibits leaf through a book and pause to look

Table 11.4. **Percentage of Selected Subgroups in Competitive Jobs at Month 22**[a]

Subgroup	Experimental group mean	Control group mean	Difference
Level of retardation			
Moderate	39	11	28**
Mild	28	16	12**
Borderline	34	29	5
Age at enrollment			
Younger than 22	30	22	8
22 or older	32	12	20**
Receipt of public assistance at enrollment			
SSI/SSDI	28	15	13*
Other public assistance	42	16	26**
No public assistance	23	26	−3
Gender			
Male	35	18	17**
Female	25	20	5
Causes of retardation			
Organic	34	10	24**
Nonorganic	30	21	9**
Work experience in 2 years prior to enrollment			
Competitive job lasting 3 months or more	52	21	31**
Other job lasting 3 months or more	28	31	−3
No job lasting 3 months	27	11	16**

[a]Adapted from Kerachsky et al. (1985).
*Statistically significant at the 10% level.
**Statistically significant at the 5% level.

at the tables and graphs rather than read the text. The implication for your analysis report is obvious.

   d. Use of Figures.   The old adage "one picture is worth a thousand words" is all too true. Note how easy it is to interpret the data from Kerachsky *et al.* (1985), shown in Figure 11.1. With a very quick glance, you can see differences in the types of jobs held by experimental and control subjects, compared to their baseline condition. The figure presents the big picture; the text can then be used to discuss the results and their statistical significance.

   Again, the manual of the American Psychological Association presents a number of criteria for the use of a figure, including simplicity, clarity, and continuity. These are presented in the bottom section of Table 11.5. In addition

Table 11.5. **Criteria regarding the Use of Tables and Figures**[a]

Tables

  1. Is the table necessary? If you discuss every item of the table in the text, the table is unnecessary.
  2. Are all comparable tables in the report consistent in presentation?
  3. Is the title brief but explanatory?
  4. Does every column have a column heading?
  5. Are all abbreviations, underlines, parentheses, and special symbols explained?
  6. Are all probability level values correctly identified, and are asterisks attached to the appropriate table entries?
  7. Are the notes in the following order: general note, specific note, probability level note?
  8. Is the table referred to in the text?
  9. Are there too many facts given in the table to permit easy interpretation?

Figures

  1. Augments rather than duplicates the text.
  2. Conveys only essential facts.
  3. Omits visually distracting detail.
  4. Is easy to read.
  5. Is easy to understand.
  6. Its purpose is readily apparent.
  7. Is consistent with and is prepared in the same style throughout the report.
  8. Is carefully planned and prepared.

[a]Adapted from the *Publication Manual of the American Psychological Association* (1984, pp. 83–95).

to these criteria, the interested reader might want to follow suggestions on graphic displays such as those found in Tufte (1983).

V. Communicating from a Marketing Perspective

Despite many people's wishes, translating program analysis results into action doesn't just happen. Organizational change literature suggests that the probability of an organization changing rests upon the following marketing factors (Beer, 1980), which promise that the change

- comes closer than any other choice in satisfying the most people
- is faster than other options
- results in immediate and long-term performance improvement
- is acceptably disruptive to the organization
- is internalized by people throughout the organization
- is supported by critical organization members
- is operationalized through changed policies and procedures

Thus, promoting change based upon the program analysis frequently requires

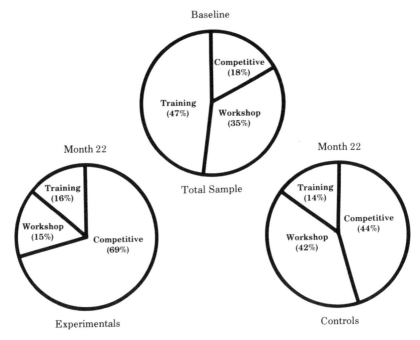

Figure 11.1. Trends in types of jobs held by working experimentals and controls (adapted from Kerachsky *et al.*, 1985).

using marketing techniques and strategies. This marketing process is facilitated through maximizing the following four basic marketing factors:

- *Product:* having what customers want or need
- *Price:* making it available at the right price (defined broadly to include incentives)
- *Placement:* offering it where, how, and when it is wanted
- *Promotion:* improving image and awareness

Because evaluation producers and consumers must market an evaluation's results in light of the target audience and contextual variables, we now discuss how to use these four marketing factors.

1. Product

The product is not just the program analysis as reflected in the final report or document, but it is also the credibility, sensitivity, and attractiveness factors discussed in the previous sections. Therefore, one should keep in mind the importance of how the analysis is "packaged" as one attempts to communicate the findings. Packaging variables include the title (neat and clean); the cover (unobstructed); the executive summary (brief, logical, and forward); use of ample tables and graphs, overviews, and summaries to each section; writing style; neat typing or word-processing; and judicious use of references to the literature (numerous references both are distracting and raise the issue of originality).

2. Incentives (Price)

Who are the consumers of program analysis? In Chapter 1 we proposed at least three groups whose perspectives on incentives and program analysis vary greatly. The three include the participant, the agency, and the larger system.

Participant. Participants, family, and advocacy groups want good services. From their perspective, program analysis should improve the quality and availability of appropriate services. Thus, communicating analysis results that stress potential programmatic improvements will increase the incentive for program analysis and implementation from the participant's perspective.

Agency. It is a safe assumption that agencies and their personnel want to provide quality services to their clientele. To the agency, program analysis should provide feedback regarding the program's effectiveness and efficiency and whether it is meeting its goals and objectives. The major incentive to the agency is to improve its effectiveness and efficiency.

System. The incentives to the system primarily concern issues of efficiency and equity. As we discussed in Chapters 1 and 9, administrators and society who are making decisions about human service policy and programs seek to find programs that are efficient and equitable. Efficient programs are those that serve to increase the net value of the goods and services available to society. Equitable programs contribute to balancing the needs and desires of the various groups in society, which is particularly important since a goal of many such programs is to increase social equity by reallocating resources or equalizing opportunities.

It is when you view program analysis and its implementation from the perspectives of the participant, agency, and system that you better understand the complexities and difficulties involved in communicating your results. No constituent group wants to play a no-win game; all have vested interests and needs. Thus, you must be sensitive to the incentives that accrue to the various constituencies when attempting to translate program analysis results into action and focus your marketing strategies accordingly.

3. Placement

The placement of program analysis refers to providing it where, how, and when it is wanted. The timing of analysis and its implementation is very important, since organizations are more amenable to analysis and potential change at some times than they are at others. In that sense, the analyst's role is much like that of a detective and prosecuting attorney (Wiseman, 1974). As a detective, the analyst should be sensitive to receptive periods as reflected in one or more of the following:

- *Opportunities* provided through internal or external program evaluation/analysis initiatives.
- *Formative evaluation studies* done by the agency, suggesting either high or low output with a corresponding request to know how the program can be improved.
- *Requirements* for program evaluation, such as accreditation, licensure, certification, or funding.
- *Requests* from agencies to analyze their program in regard to process, impact, and/or benefit–cost analysis and to possibly make programmatic changes.

As a prosecuting attorney, the analyst may need to convince a jury of skeptical readers that the time is appropriate for the proposed changes. In that sense, the following placement principles proposed by Glaser (1973) are appropriate:

1. *Credibility,* which stems from the soundness of evidence for the analysis's value or from its espousal by highly respected persons or institutions.
2. *Observability,* or the opportunity to see a demonstration of the innovation or its results in practice.
3. *Relevance* to coping with (or solving) a problem persistent and sharply bothersome to decision makers or influential members of the constituency.
4. *Relative advantage* over current practices or a perception that the change will more than offset the effort involved in the change.
5. *Implementation ease,* as opposed to difficulty, in operationalizing the analysis or in transplanting it to different settings.
6. *Compatibility* with the program constituents' established values, norms, and procedures.
7. *Testability,* which permits a pilot study and therefore does not require an irreversible commitment by the system.

We have found that translating program analysis results into action requires addressing these issues in the public policy section of the final report. We additionally have found that among the seven principles listed, skeptical decision makers are frequently most amenable to those principles related to observability, implementation ease, and testability.

4. Promotion

The previous three sections on the marketing principles of product, price (incentive), and placement suggest that translating program analysis results into action requires someone to promote the analysis and its accompanying findings. The promoter is typically the administrator or decision maker who is responsible for implementing the results of the program analysis through reducing organizational resistance, dissipating political opposition, facilitating the necessary organizational changes and channeling the program into new directions. How this is done depends upon a number of factors, but we suggest the following principles, based upon our experiences as—and with—promoters:

1. Effective adoption of innovation is more likely to occur if it is planned and implemented using all levels of personnel.
2. Effective implementation is more likely if manuals, protocols, or blueprints are available and used.
3. Successful implementation is more likely if the process is subject to data guidance, observation, and feedback.

VI. Summary

Program administrators are constantly making decisions about day-to-day program operations and possible programmatic changes. They are busy people who do not have sufficient time to do all that is required. Thus, they make decisions quickly and may not follow the ideal decision-making process of diagnosing, evaluating all the alternatives, and choosing the best alternative based on the available facts. Rather, a more realistic decision-making process might include the following (Alexis & Wilson, 1967): (1) search or information-gathering strategies are used that are directly related to the ''cognitive strain'' imposed by time, information retention, and recall; (2) since problem-solving behavior is adaptive, an administrator starts with a tentative solution, searches for information, modifies the initial solution, and continues the activity until there is some balance between expected and realized behaviors; and (3) an administrator makes decisions based largely on facts that are easily assimilated, on preferences, or on aversion to risk.

Thus, simply conducting program analysis is not sufficient to impact the administrator's decision-making process. In addition to the analysis, you need to communicate the evaluation results using clear and effective techniques. This chapter has summarized four techniques that we feel will assist you in communicating your findings and, consequently, impact the administrator's or policymaker's decision-making process. The four techniques were establishing your credibility, being sensitive to the needs of the target audience and the context within which that audience operates, presenting your results in an attractive, concise, and readable format, and communicating from a marketing perspective.

The 1970s and 1980s have seen a huge growth in the number of human service program evaluations. An administrator trying to sort through these evaluations is often faced with the task of determining not only the relative effectiveness of the various programs but also the relative accuracy of the evaluations. We hope the suggestions we have made about credibility, presentation style, and marketing perspective will greatly assist the administrator in deciding on the utility and potential of the program analysis presented.

VII. Additional Readings

Bennis, W. G., Benne, K. D., Chin, R., & Corey, K. E. (Eds.). (1976). *The planning of change.* New York: Holt, Rinehart & Winston.

Buchanan, D. A., & Boddy, D. (1983). *Organizations in the computer age.* Brookfield, VT: Gower Publication Company.

Davis, H. R., & Salasin, S. E. (1975). The utilization of evaluation. In E. L. Struening & M. Guttentag (Eds.), *Handbook of evaluation research, Vol. 1* (pp. 621–666). Beverly Hills, CA: Sage.

Fairweather, G. W. (1977). A process of innovation and dissemination experimentation. In L. Rutman (Ed.), *Evaluation research methods: A basic guide* (pp. 179–196). Beverly Hills, CA: Sage.

Gottfredson, G. D. (1984). A theory-ridden approach to program evaluation: A method for stimulating researcher–implementer collaboration. *American Psychologist, 39*(10), 1101–1112.

Mitchell, D. E., & Spady, W. F. (1978). Organizational contexts for implementing outcome evaluation. *Evaluational Researcher, 7,* 9–17.

How Am I Doing?

After more and more changes, we are
more or less the same.
(Simon and Garfunkel's "The Boxer")

I. Overview

We hope the above song line is not true, for we have attempted throughout the book to present both ideas and techniques to improve your ability to be a better producer or consumer of program evaluation. We have suggested that program evaluation is an untraveled road, rocky and full of quagmires. Having a field guide will help you avoid those areas where you need fancy equipment; but there are many areas one can go and enjoy tremendously by simply following the map (field guide) that we have proposed.

In developing this field guide, we were acutely aware that administrators are constantly having to make program choices and day-to-day decisions, along with providing services. Thus, we have tried to increase the efficiency of the time administrators devote to evaluation since, as the Chinese say, "efficiency is time, and time is life." To increase the administrator's efficiency of time, we have presented some rules of evidence, have tried to focus your efforts on questions that can be answered by available resources, and have outlined criteria that you can use to quickly evaluate the issues of an evaluation study. But we have also stressed that the suggested guidelines and broad brush approach do not replace highly structured and systematic program analyses. Rather, the rules and approaches presented in the field guide, even though they contain some error, will help you make the best use of your time.

But now it's time to ask, how am I doing? Throughout the book we have asked you to complete various table shells and self-surveys to help you think through the issues being discussed and to evaluate your current status. We finalize that process in this chapter and provide you with a brief review of the major themes that have appeared throughout the book. These are the general principles that we hope you incorporate into your mind-set regarding program evaluation.

II. The Book's Themes

Throughout the book a number of themes have been presented reflecting the authors' approach to program evaluation. The more important of these are listed in Table 12.1. These themes underlie the rules of thumb, techniques, guidelines, and cautions presented throughout the book. In a very real sense, they also reflect our motive for writing the book: to help administrators match resources to evaluation questions asked and to make better decisions. As we discussed in the preface, there is frequently a misallocation of evaluation resources, wherein the

---

Table 12.1. **Common Themes Appearing throughout the Book**

1. Analysts should be systematic in their collection and analysis of program data; their conclusions should be based on thoughtful, correct, and honest analysis. They should use logical thought and follow the scientific method.
2. Administrators are constantly faced with having to make decisions that involve incomplete information and some uncertainty; therefore, decisions must be based on experience and value judgments in addition to good evaluation information.
3. Program analysis provides rules of evidence for organizing information to make judgments.
4. No human service program occurs in isolation; rather, it is a part of a larger environment that has specific economic, social, and political characteristics.
5. Evaluation is undertaken to facilitate decision making, and decisions are choices. Hence, evaluation involves structured comparisons to determine what happens under various choices. Analysis techniques cannot make decisions; they can only help smooth and clarify the decision process.
6. Administrators are both evaluation producers and consumers. As producers, they want to: carefully define the questions they want to answer; collect information about those questions; interpret the information, making sure their conclusions follow logically from what they observe. The major question for consumers is whether there is enough certainty and applicability to their situation to act on the evaluation's results.
7. Evaluation questions should be matched to available resources.
8. Administrators and society need to find programs that are efficient (serve to increase the net value of the goods and services available to society) and equitable (contribute to balancing the needs and desires of the various groups in society).
9. Program analysis should be an integral part of an agency or program. Regardless of a program's size or level of sophistication, some degree of program analysis can be conducted.
10. Program analysis results are more likely to be accepted if they are presented clearly and honestly. Evaluation involves clear, logical thinking.

---

program does not have adequate resources to answer large-scale evaluation questions frequently asked of it. Additionally, during the 1970s and 1980s we have seen a huge increase in the number of human service evaluations that taxes the administrator to determine the relative effectiveness of the various programs and the relative accuracy of the evaluations. Thus, the techniques that we have presented, based on the themes listed in Table 12.1, will help guide the administrator in choosing which questions to answer if resources are limited. They will also provide quick access to information about analytical techniques and interpretive guidelines.

III. Evaluating ''How Am I Doing?''

There are a number of ways to answer the question, how am I doing? Having read the book you may subjectively feel more comfortable with program evaluation. By following the suggestions we've made throughout the book you might feel that you:

- are a better producer or consumer of program evaluation
- have a better perspective regarding program evaluation
- are better able to do some level of process, impact, or benefit–cost analysis
- can effect programmatic change by implementing program evaluation results

We hope that you have these feelings, and if you do, you're headed in the right direction. However, we suggest that a more objective evaluation also is appropriate. Such an evaluation is outlined in Survey 12.1. The survey is directed more at the program producer, but it can also be used by the consumer in evaluating others' program evaluations. The survey questions focus on how you are doing in reference to either the rules of thumb governing program evaluation or the various components of process, impact, or benefit–cost analysis. We do not mean to imply that the two sections of the survey are discrete, but we feel that by separating the survey into suggested guidelines and types of analysis, you can determine how you are doing either generally or in reference to a specific type of analysis.

As with Survey 11.1, we are not proposing a passing or failing grade. We do suggest, however, that your ability as an evaluation producer or consumer increases as the number of ''yes'' answers increase (a perfect score is 25, by the way). For those questions on which your answer is no, you might want to refer to the flowchart presented in Figure 3.2 to find relevant parts of the book wherein the item is discussed.

Survey 12.1. **Self-Survey regarding "How Am I Doing?"**

How am I doing in reference to the guidelines governing program evaluation	Evaluation[a]		How am I doing in reference to attempted process, impact, or benefit–cost analysis	Evaluation[a]	
Analysis component	Yes	No	Analysis type	Yes	No
1. In reference to the *setup*, can you describe the:			1. In reference to *process analysis*, have you included:		
a. problem addressed			a. a description of the persons		
b. persons served			b. a description of the program, including internal and external factors		
c. services provided			c. cost estimates such as costs per participant per fiscal year		
d. evaluation context			2. In reference to *impact analysis*, have you included:		
e. expected outcomes			a. measures of program outcomes		
f. rationale linking the intervention/services to the outcome(s)			b. impact estimates		
2. In reference to *marshalling the evidence*, have you:					
a. defined the outcomes					
b. selected measures that reflect					

initial intent

c. described the program's geographical context

d. measured the services delivered

e. measured changes in outcomes

f. collected data on both sides of the structured comparison

3. In reference to *interpreting the findings*, have you:

a. evaluated the correspondence between the setup and marshalling the evidence

b. attributed change in outcomes to the intervention

c. interpreted conclusions within the program's context

d. addressed the issue of uncertainty

c. reference to the uncertainty surrounding the estimates

3. In reference to *benefit–cost analysis*, have you:

a. developed the accounting framework

b. estimated benefits and costs

c. presented and interpreted the results

[a]Place a check beneath either Yes or No.

Your success with program evaluation is a function of many things, including time, resources, and expertise. We hope the information in this book has increased your knowledge of, and interest in, program evaluation. If it has, you might want to refer to the references we have listed at the end of each chapter for additional information. In addition, you might also:

- Call or write organizations such as The Association for Public Policy and Management (c/o Institute of Policy Sciences, Duke University, 4875 Duke Station, Durham, NC 27706; 919-684-6612).
- Contact sources at your local university or college.
- Talk to someone in the field in which you are interested, including persons in such academic departments as applied economics, public policy, criminology, health care, human services, public health, and program evaluation.
- Access currently available program evaluations through computerized literature searches (such as ERIC or Medline/Medlars) or the federal government's clearing house.

As we have stated repeatedly, all programs are amenable to some type of program analysis. Regardless of how you scored on the self-survey, we hope that you will be bold and proceed with one or more of the types of program analysis we have presented in the field guide. If you do, then you will be able to tell others how you are doing and will no longer need to ask someone else.

In the end, administrators make decisions that reflect their assessment of the political and social implications of programs as well as the objective evidence about program impacts. Administrators will never have complete information and certainty; thus, decisions must be based on experience and value judgments in addition to evaluation information. The real benefit of those analytic procedures we have presented is that they help you organize and interpret available data. Analytic techniques cannot make decisions; they can only help smooth and clarify the decision process.

Thus, the book ends where it began, by stressing the critical need in the late 1980s and 1990s for program analysis activities such as those presented. Research into the effectiveness of human service programs has a mixed record, at best. A few studies have shown the potential of program analysis to help organize the available resources to best serve the needs of society. Nevertheless, the potential of most studies is not realized because of flawed methods and inadequate observation, overstatement of conclusions, or unintelligibly presented results. We hope that the impact of the book will be to improve this situation by increasing the awareness, use, and quality of program analysis in the human services field. If that occurs, our efforts will be well rewarded.

References

Alexander, J. F., & Parsons, B. V. (1973). Short-term behavioral interventions with delinquent families. *Journal of Abnormal Psychology, 81*(2), 219–225.

Alexis, M., & Wilson, C. Z. (1967). *Organizational decision-making.* Englewood Cliffs, NJ: Prentice-Hall.

American Psychological Association (1984). *Publication manual of the American Psychological Association* (3rd ed.). Washington, DC: Author.

Andrews, F. R., & Withey, S. B. (1976). *Social indicators of well-being: American's perceptions of life quality.* New York: Plenum.

Aninger, M., & Bolinsky, K. (1977). Levels of independent functioning of retarded adults in apartments. *Mental Retardation, 15*(4), 12–13.

Anthony, W. A., & Farkas, M. (1982). A client outcome planning model for assessing psychiatric rehabilitation interventions. *Schizophrenia Bulletin, 8*(1), 13–38.

Ashbaugh, J., & Allard, M. A. (1983). *Longitudinal study of the court-ordered deinstitutionalization of Pennhurst residents: Comparative analysis of the costs of residential and day services within institutional and community settings.* Boston: Human Services Research Institute.

Attkisson, C. C., Hargreaves, W. A., Horowitz, M. J., & Sorensen, L. E. (Eds.). (1978). *Evaluation of human service programs.* New York: Academic Press.

Baker, B., Brightman, A., & Hinchaw, S. (1980). *Toward independent living.* Champaign, IL: Research Press.

Baker, F., & Intagliata, J. (1982). Quality of life in the evaluation of community support systems. *Evaluation and Program Planning, 5,* 69–79.

Beer, M. (1980). *Organization change and development: A systems review.* Santa Monica, CA: Goodyear.

Bellamy, G. T., Rhodes, L., & Albin, J. M. (1985). Supported employment. In W. Kiernan & J. Stark (Eds.), *Pathways to employment for developmentally disabled adults* (pp. 129–138). Baltimore, MD: Paul H. Brookes.

Bloomenthal, A. M., Jackson, R., Kerachsky, S., Stephens, S., Thornton, C., & Zeldis, K. (1982). *SW/STETS evaluation: Analysis of alternative data-collection strategies.* Princeton, NJ: Mathematica Policy Research.

Brown, M., Diller, L., Gordon, W. A., Fordyce, W. E., & Jacobs, D. F. (1984). Rehabilitation indicators and program evaluation. *Rehabilitation Psychology, 29*(1), 21–35.

Bryson, J. M., & Cullen, J. W. (1985). A contingent approach to strategy and tactics in formative and summative evaluations. *Evaluation and Program Planning, 7,* 267–290.

Campbell, A., Converse, P. E., & Rogers, W. L. (1976). *The quality of American life: Perceptions, evaluations and satisfactions.* New York: Russell Sage.

Campbell, D. T., & Erlebacher, A. (1975). How regression artifacts in quasi-experimental evaluations can mistakenly make compensatory education look harmful. In E. Struening & M. Guttentag (Eds.), *Handbook of evaluation research, Vol. 1* (pp. 597–620). Beverly Hills, CA: Sage.

Carcagno, G. J., Applebaum, R., Christianson, J., Phillips, B., Thornton, C., & Will, J. (1986).

The evaluation of the national long-term care demonstration: The planning and operational experience of the channeling projects: Volumes 1 and 2. Princeton, NJ: Mathematica Policy Research.

Carter, D. E., & Newman, F. L. (1976). *A client-oriented system of mental health service delivery and program management: A workbook and guide.* Rockville, MD: National Institute of Mental Health.

Ciarlo, J. A., Brown, T. R., Edwards, D. W., Kiresuk, T. J., & Newman, F. L. (1986). *Assessing mental health treatment outcome measurement techniques.* DHHS Publication No. ADM 86-1301. Washington, DC: U.S. Government Printing Office.

Comfort, L. K. (1982). *Educational policy and evaluation: A context for change.* New York: Pergamon Press.

Conley, R. W., Noble, J. H., Jr., & Elder, J. K. (1986). Problems with the service system. In W. E. Kiernan & J. A. Stark (Eds.), *Pathways to employment for developmentally disabled adults* (pp. 67–84). Baltimore: Paul H. Brookes.

Cooley, W. W., & Bickel, W. E. (1986). *Decision-oriented educational research.* Hingham, MA: Kluwer Academic Publishers.

Cronbach, K. J., & Furby, L. (1970). How we should measure "change"—or should we? *Psychological Bulletin, 74,* 68–80.

Davidson, W. S., Koch, J. R., Lewis, R. S., & Wresinski, R. (1981). *Evaluation strategies in criminal justice.* New York: Pergamon Press.

Dearman, N. B., & Plisko, V. W. (1982). The condition of vocational education: 1982 edition. Washington, DC: U.S. Department of Education, National Center for Education Statistics.

Demone, H. W., Jr., & Harshbarger, D. (1973). *The planning and administration of human services.* New York: Behavioral Publications.

Dolbeare, K. M. (Ed.). (1975). *Public policy evaluation.* Beverly Hills, CA: Sage.

Edgerton, R. B. (1975). Issues relating to quality of life among mentally retarded individuals. In M. J. Begab & S. A. Richardson (Eds.), *The mentally retarded and society: A social service perspective* (pp. 185–198). Baltimore, MD: University Park Press.

Edwards, W., Guttentag, M., & Snapper, K. (1975). A decision-theoretic approach to evaluation research. In E. L. Struening & M. Guttentag (Eds.), *Handbook of evaluation research, Vol. 1* (pp. 139–182). Beverly Hills, CA: Sage.

Fairweather, G. W., & Davidson, W. S. (1986). *An introduction to community experimentation: Theory, methods and practice.* New York: McGraw-Hill.

Firestone, W. A., & Herriott, R. E. (1983). The formalization of qualitative research: An adaptation of "soft" science for the policy world. *Evaluation Review, 17*(4), 437–466.

Flanagan, J. C. (1978). A research approach to improving our quality of life. *American Psychologist, 33,* 138–147.

Gersten, R., Crowell, F., & Bellamy, T. (1986). Spillover effects: Impact of vocational training on the lives of severely mentally retarded clients. *American Journal of Mental Deficiency, 90*(5), 501–506.

Glaser, E. (1973). Knowledge transfer and institutional change. *Professional Psychology, 10,* 434–444.

Gramlich, E. M. (1981). *Benefit–cost analysis of government programs.* Englewood Cliffs, NJ: Prentice-Hall.

Greenberg, D. A., & Robins, P. K. (1986). The changing role of social experiments in policy analysis. *Journal of Policy Analysis and Management, 5*(2), 340–362.

Greene, D., & David, J. L. (1984). A research design for generalizing from multiple case studies. *Evaluation and Program Planning, 7*(1), 73–85.

Hagedorn, H. J., Beck, K. J., Neubert, S. F., & Welin, S. H. (1979). *A working manual of simple program evaluation techniques for community mental health centers.* Rockville, MD: National Institute of Mental Health.

Hall, R. E. (1979). Comments on the labor market displacement effect in the analysis of the net impact of manpower training programs by George Johnson. In F. E. Bloch (Ed.), *Evaluating manpower training programs* (pp. 255–258). Greenwich, CT: JAI Press.

Hanson, E. M. (1979). School management and contingency theory: An emerging perspective. *Educational Administration Quarterly, 15,* 98–116.

Hanushek, E. (1986). The economics of schooling: Production and efficiency in the public schools. *Journal of Economic Literature, 25*(3), 1141–1177.

Hauber, F., Bruininks, R., Hill, B., Lakin, K. C., and White, C. (1984). National census of residential facilities: Fiscal year 1982. Minneapolis: Department of Educational Psychology, University of Minnesota.

Haveman, R. H., & Wolfe, B. L. (1984). Schooling and economic well-being: The role of non-market effects. *Journal of Human Resources, 19*(3), 377–407.

Havlock, R. G. (1970). *A guide to innovation in education.* Ann Arbor: Center for Research on Utilization of Scientific Knowledge, Institute for Social Research, University of Michigan.

Heal, L. W., & Fujiura, G. T. (1984). Methodological considerations in research on residential alternatives for developmentally disabled persons. In N. R. Ellis & N. W. Bray (Eds.), *International review of research in mental retardation, Vol. 12* (pp. 206–244). New York: Academic Press.

Hellreigel, D., & Slocum, J. W., Jr. (1982). *Management: Contingency approached* (3rd ed.). Reading, MA: Addison-Wesley.

Hill, M., Hill, J. W., Wehman, P., & Banks, P. D. (1985). An analysis of monetary and nonmonetary outcomes associated with competitive employment of mentally retarded persons. In P. Wehman & J. W. Hill (Eds.), *Competitive employment for persons with mental retardation: From research to practice* (pp. 85–102). Richmond, VA: Rehabilitation Research and Training Center, School of Education, Virginia Commonwealth University.

Hill, M., & Wehman, P. (1985). Cost–benefit analysis of placing moderately and severely handicapped persons into competitive work. *Journal of the Association for the Severely Handicapped, 8*(1), 30–38.

Hollister, R., Kemper, P., & Maynard, R. (1984). *The national supported work demonstration.* Madison: University of Wisconsin Press.

Johnson, G. E. (1979). The labor market displacement effect in the analysis of the net impact of manpower training programs. In F. E. Bloch (Ed.), *Evaluating manpower training programs* (pp. 227–254). Greenwich, CT: JAI Press.

Jones, D. D. (1986). The treatment of fringe benefits in personal injury and wrongful death loss calculations. *Benefits Quarterly, 2*(4), 59–62.

Kakalik, J. S., Furry, W. S., & Thomas, M. A. (1981). *The costs of special education: Summary of study findings.* Santa Monica, CA: Rand Corporation.

Kemper, P. (Ed.). (1986). The evaluation of the national long-term care demonstration: Final report. Princeton, NJ: Mathematica Policy Research.

Kemper, P., & Long, D. A. (1981). *The supported work evaluation: Technical report on value of in-program output and costs.* Princeton, NJ: Mathematica Policy Research.

Kemper, P., Long, D. A., & Thornton, C. (1981). *The supported work evaluation: Final benefit–cost analysis.* Princeton, NJ: Mathematica Policy Research.

Kerachsky, S., Thornton, C., Bloomenthal, A., Maynard, R., & Stephens, S. (1985). *Impacts of transitional employment for mentally retarded young adults: Results of the STETS demonstration.* Princeton, NJ: Mathematica Policy Research.

Lambert, M. J., Christensen, E. R., & DeJulio, S. S. (1983). *The assessment of psychotherapy outcomes.* New York: Wiley.

Landesman-Dwyer, S. (1981). Living in the community. *American Journal of Mental Deficiency, 86*(3), 223–234.

Lansing, J. B., & Morgan, J. N. (1971). *Economic survey methods.* Ann Arbor: Institute for Social Research, University of Michigan Press.

Lerman, P. (1975). *Community treatment and social control.* Chicago: University of Chicago Press.

Lewis, D. R., Bruininks, R. H., & Thurlow, M. (1986). Cost analysis for district level special education planning, budgeting and administrating. Minneapolis: University of Minnesota, Department of Psychoeducational Studies.

Long, D. A., Mallar, C. D., & Thornton, C. V. D. (1981). Evaluating the benefits and costs of the Job Corps. *Journal of Policy Analysis and Management, 1*(1), 55–76.

Luborsky, L., & Backrach, H. M. (1974). Factors influencing clinicians' judgments of mental health: Experiences with health–sickness rating scale. *Archives of General Psychiatry, 31,* 292–299.

Mallar, C. D., & Thornton, C. (1978). Transitional aid for released prisoners: Evidence from the LIFE experiment. *Journal of Human Resources, 13*(2), 115–140.

Mallar, C. D., & Thornton, C. (1980). *Benefit–cost methodology for studying apprenticeship programs.* Princeton, NJ: Mathematica Policy Research.

Mank, D. M., Rhodes, L. E., & Bellamy, G. T. (1986). Four supported employment alternatives. In W. E. Kiernan & J. A. Stark (Eds.), *Pathways to employment for adults with developmental disabilities* (pp. 139–153). Baltimore, MD: Paul H. Brookes.

Mayeda, T., & Lindberg, G. (1980). *Performance measures of skill and adaptive competencies.* Pomona, CA: Lanterman State Hospital and Developmental Center.

McCloskey, D. (1985). Economic writing. *Economic Inquiry, 24,* 187–198.

McDill, E. L., McDill, M. S., & Spreke, J. T. (1972). Evaluation in practice: Compensatory education. In P. H. Rossi and W. Williams (Eds.), *Evaluating social programs: Theory, practice and politics* (pp. 141–186). New York: Seminar Press.

Meyers, C. E., Nihira, K., & Zetlin, A. (1979). The measurement of adaptive behavior. In N. Ellis (Ed.), *Handbook of mental deficiency* (2nd ed.) (pp. 150–164). Hillsdale, NJ: Lawrence Erlbaum.

Murphy, J. T. (1980). *Getting the facts: A field work guide for evaluators and policy analysts.* Santa Monica, CA: Goodyear.

National Institute of Mental Health (1973). *Planning for creative change in mental health services: A distillation of principles in research utilization, Volumes 1 and 2.* Washington, DC: U.S. Government Printing Office.

Newman, F. L., & Howard, K. I. (1986). Therapeutic effort, treatment outcome and national health policy. *American Psychologist, 41*(2), 181–187.

Nunnally, J. C. (1975). The study of change in evaluation research: Principles concerning measurement, experimental design and analysis. In E. Struening & M. Guttentag (Eds.), *Handbook of evaluation research, Vol. 1* (pp. 101–138). Beverly Hills, CA: Sage.

O'Brien, J. (Ed.). (1976). *Evaluative research on social programs for the elderly.* Washington, DC: U.S. Department of Health, Education and Welfare (Publ. 77-20120).

Pechman, J. A. (1985). *Who paid the taxes, 1966–1985.* Washington, DC: The Brookings Institute, Studio of Government Finance.

Poincaré, J. H. (1968). *La science et l'hypothèse.* Paris, France: Flammarion.

Popham, W. J. (1975). *Educational evaluation.* Englewood Cliffs, NJ: Prentice-Hall.

Posavac, E. J., & Carey, R. G. (1980). *Program evaluation: Methods and case studies.* Englewood Cliffs, NJ: Prentice-Hall.

Pruchno, R. A., Boswell, P. C., Wolff, D. S., & Foletti, M. V. (1983). A community mental health program: Evaluating outcomes. In M. A. Smyer & M. Gatz (Eds.), *Mental health and aging: Programs and evaluations* (pp. 72–87). Beverly Hills, CA: Sage.

Rusch, F., & Mithaug, D. (1980). *Vocational training for mentally retarded adults: A behavioral analytic approach.* Champaign, IL: Research Press.

Schalock, R. L. (1983). *Service for the developmentally disabled adult: Development, implementation and evaluation.* Austin, TX: Pro-ed.

Schalock, R. L. (1985). Comprehensive community services: A plea for interagency collaboration. In R. H. Bruininks & K. C. Lakin (Eds.), *Living and learning in the least restrictive environment* (pp. 37–64). Baltimore, MD: Paul H. Brookes.

Schalock, R. L., Gadwood, L. S., & Perry, P. B. (1984). Effects of different training environments on the acquisition of community living skills. *Applied Research in Mental Retardation, 5,* 425–438.

Schalock, R. L., Harper, R. S., & Genung, T. (1981). Community integration of mentally retarded adults: Community placement and program success. *American Journal of Mental Deficiency, 85*(5), 478–488.

Schalock, R. L., & Hill, M. (1986). Evaluating employment services. In W. Kiernan & J. Stark (Eds.), *Pathways to employment* (pp. 285–302). Baltimore, MD: Paul H. Brookes.

Schalock, R. L., & Keith, K. D. (1986). Resource allocation approach for determining clients' need status. *Mental Retardation, 24*(1), 27–35.

Schalock, R. L., Keith, K. D., Karan, O. C., & Hoffman, K. (in press). Quality of life measurement: Its use as an outcome measure in human service programs. *Mental Retardation.*

Schalock, R. L., & Lilley, M. A. (1986). Placement from community-based mental retardation programs: How well do clients do after 8–10 years? *American Journal of Mental Deficiency, 90*(6), 669–676.

Schalock, R. L., Wolzen, B., Ross, I., Elliott, B., Werbel, G., & Peterson, K. (1986). Post-secondary community placement of handicapped students: A five-year follow-up. *Learning Disability Quarterly, 9,* 295–303.

Schauss, A. G., & Simonsen, C. E. (1979). A critical analysis of the diets of chronic juvenile offenders. Part I. *Orthomolecular Psychiatry, 8*(3), 149–157.

Schneider, L. C., & Struening, E. L. (1983). SLOF: A behavioral rating scale for assessing the mentally ill. *Social Work Research and Abstracts, 1,* 9–21.

Shortell, S. M., Wickizer, T. M. & Wheeler, R. G. (1984). *Hospital–physician joint ventures: Results and lessons from a national demonstration in primary care.* Ann Arbor, MI: Hiatt Administration Press.

Social Security Administration. (1985). *Social security bulletin: Annual statistical supplement, 1984–85.* Washington, DC: Author.

Stainback, S., & Stainback, W. (1984). Methodological considerations in qualitative research. *Journal of the Association for Persons with Severe Handicaps. 9*(4), 296–303.

Strunk, W., Jr., & White, E. B. (1979). *The elements of style* (3rd ed.). New York: Macmillan.

Thornton, C. (1981). *The benefits and costs of SW-STETS: A design overview.* Princeton, NJ: Mathematica Policy Research.

Thornton, C. (1984). Benefit–cost analysis of social programs. In R. H. Bruininks & C. K. Lakin (Eds.), *Living and learning in the least restrictive environment* (pp. 225–244). Baltimore, MD: Paul H. Brookes.

Thornton, C., & Dunstan, S. M. (1986). *The evaluation of the national long-term care demonstration: Analysis of the benefits and costs of channeling.* Princeton, NJ: Mathematica Policy Research.

Thornton, C., Long, D., & Mallar, C. (1982). *A comparative evaluation of Job Corps after forty-eight months of postprogram observation.* Princeton, NJ: Mathematica Policy Research.

Thornton, C., Will, J., & Davies, M. (1986). *The evaluation of the national long-term care demonstration: Analysis of channeling project costs.* Princeton, NJ: Mathematica Policy Research.

Tufte, E. R. (1983). *The visual display of quantitative information.* Cheshire, CT: Graphics Press.

U.S. Chamber of Commerce. (1986). *Employee benefits, 1985.* Washington, DC: Research Center, Economic Policy Division, U.S. Chamber of Commerce.

U.S. Department of Labor. (1977). *A nationwide report on sheltered workshops and their employment of handicapped individuals. Volume I: Workshop survey.* Washington, DC: U.S. Government Printing Office.

U.S. Department of Labor. (1980). *Employee compensation in the private nonfarm economy, 1977* (Summary 80-5). Washington, DC: Bureau of Labor Statistics.

U.S. Office of Management and Budget (OMB) (1972). *Discount rates to be used in evaluating time-distributed costs and benefits.* (OMB Circular No. A-94). Washington, DC: OMB.

Vera Institute of Justice and Job Path (1983). *A report to the Helena Rubenstein Foundation, Inc.* New York: Vera Institute of Justice.

Warner, K. E., & Luce, B. R. (1982). *Cost–benefit and cost-effectiveness analysis in health care.* Ann Arbor, MI: Health Administration Press.

Washow, I. W., & Parloff, M. B. (Eds.). (1975). *Psychotherapy change measures.* Washington, DC: U.S. Department of Health, Education and Welfare, National Institute of Mental Health (Publication No. ADM 74-120).

Weis, C. H. (1973). Where politics and evaluation research meet. *Evaluation, 1,* 37–45.

Weisman, M. M. (1975). The assessment of social adjustment. *Archives of General Psychiatry, 32,* 357–365.

Williams, J. M. (1981). *Style: Ten lessons in clarity and grace.* Glenview, IL: Scott, Foresman.

Wiseman, J. P. (1974). The research analyst. *Urban Life and Culture, 3*(3), 317–328.

Wright, W. E., & Dixon, M. C. (1977). Community prevention and treatment of juvenile delinquency: A review of evaluation studies. *Journal of Research in Crime and Delinquency, 14,* 35–67.

Index